Martin C. Hirsch
Dictionary of Human Neuroanatomy

WITHDRAWN

Springer

Berlin
Heidelberg
New York
Barcelona
Hong Kong
London
Milan
Paris
Singapore
Tokyo

Martin C. Hirsch

Dictionary
of Human Neuroanatomy

Springer

Dr. Martin C. Hirsch
Scheppe Gewissegasse 28
35039 Marburg
Germany

Cover page: silverBRAIN model (www.brainmedia.de)

ISBN 3-540-66523-4   Springer-Verlag Berlin Heidelberg New York

Library-of-Congress   CIP-Data applied for

Die Deutsche Bibliothek - CIP-Einheitsaufnahme
Dictionary of human neuroanatomy / Martin C. Hirsch. - Berlin ; Heidelberg ; New York ; Barcelona ;
Hong Kong ; London ; Milan ; Paris ; Singapore ; Tokyo : Springer, 1999
ISBN 3-540-66523-4

© Springer-Verlag Berlin Heidelberg 2000
Printed in Germany

Typesetting: Kirsten Matthias, Tatjana Scheid, Heidelberg
Cover design: Erich Kirchner, Heidelberg

SPIN: 10731475   27/3136   - 5 4 3 2 1 0 Printed on acid-free paper

Preface

Whether it was during my studies or, now, when preparing lectures, during scientific work or discussions of how the brain functions - over and over again I have reached a point where I need concise and precise information on a certain structure. In the standard reference books most terms are missing, and accessing information in textbooks and other scientific works or in original papers is quite tedious.

In short: a dictionary of terms in neuroanatomy and neurofunction was needed. In the course of many years of work the *Dictionary of Human Neuroanatomy* was compiled. Our purpose in preparing the dictionary was not to collect the latest scientific discussions and results but rather to present sound and comprehensive basic information which will help demonstrate the structures in their context quickly and efficiently. It should be a help to all those who need quick information on the structures of the human brain. We would like to express our sincere gratitude to all those individuals who took part in producing the dictionary: to Dr. Lange for initiating the project and for the smooth cooperation; to Ms. Wilbertz for her efficiency in supporting production; to Dr. Harder for proofreading and valuable groundwork; to my family for their loving patience – and last but not least, to iAS GmbH (www. brainmedia.de) for the illustration on the cover page.

In closing, we would like to encourage you, the readers, to please contact us if you find that information is missing, if you find any mistakes or if you have any other suggestions or requests. All suggestions will be considered and, if possible, included in the next edition. Please mail your ideas to MartinHirsch@CompuServe.com. Thank you very much!

Marburg, September 1999   Martin C. Hirsch

Structure of the Book

Name of the structure

Category:	Where can the structure be found in the brain?
TA Latin:	Name according to the Termino-logica anatomica 1998, in Latin
TA English:	Name according to the Termino-logica anatomica 1998, in English
Importance for examination:	1 = low; 2 = medium; 3 = high
Definition	

A

A-fibers of dorsal root of the spinal nerve

Medulla spinalis 2

C and A delta fibers innervate primarily nociceptors but also thermo- and some mechanoreceptors. Conversely, A-beta and A-delta fibers innervate touch receptors of the skin.

All these fiber types enter the spinal cord via the dorsal root.

A1-A7

TA Latin: *Cellulae noradrenergicae medulae oblongatae (A1)*
TA English: *Noradrenergic cells in medulla oblongata (A1)*
→ Noradrenergic cell groups A1-A7, Locus coeruleus

A8-A10

TA Latin: *Cellulae aminergicae formationis reticularis (A8)*
TA English: *Aminergic cells in reticular formation (A8)*
→ Dopaminergic cell groups A8-A10

Abducens nerve (VI)

Nerves 3
TA Latin: *N. abducens (N.VI)*
TA English: *Abducent nerve (VI)*
Abducens nerve (VI) has a purely motor function and innervates the lateral rectus muscle of the eyeball, generating an abduction movement of the eyeball (hence the name)
Nucleus: nucleus of abducens nerve.
Skull: superior orbital fissure.
Damage to the nerve causes inversion of the ipsilateral eyeball towards the nose. This produces diplopia (double vision), increasingly so the more the two visual axes deviate from each other. Looking in the direction of the respective eye reduces the severity of the diplopia.

Accessory hemiazygos vein

Vessels 3
TA Latin: *V. hemiazygos accessoria*
TA English: *Accessory hemi-azygos vein*
Accessory hemiazygos vein and hemiazygos vein together enter the azygos vein.
The catchment area of the hemiazygos vein is the upper left intercostal spaces as well as parts of the upper left mediastinum.

Accessory motor nucleus of the trigeminal nerve

1

Cells in the immediate vicinity of the motor nucleus of the trigeminal nerve.

Accessory nerve (XI)

Nerves 3
TA Latin: *N. accessorius (N.XI)*
TA English: *Accessory nerve (XI)*
The accessory nerve has two parts:

- accessory nerve (XI), cranial roots: these fibers arise from nucleus ambiguus and innervate the pharynx and larynx muscles and course together with the vagus nerve (X).
  Skull: Foramen jugulare.
- accessory nerve (XI), spinal root: it arises from a nuclear column in the cervical cord (spinal root nucleus of accessory nerve) and innervate the sternocleidomastoid muscle and the trapezius muscle.
  Skull: Foramen magnum.
  Dysfunction of the accessory nerve (XI) results in accessory paralysis rendering it more difficult to lift the arm above shoulder level (trapezius muscle), and turning the head to the unimpaired side is possible only after having successfully contended with resistance (sternocleidomastoid muscle).

Accessory nucleus of oculomotor nerve

Mesencephalon 3
TA Latin: *Nucl. accessorius n. oculomotorii*
TA English: *Accessory nucleus of oculomotor nerve*
The accessory nucleus (Edinger-Westphal) is the parasympathetic nuclear component of the oculomotor nucleus.
It contains the somas of the preganglionic parasympathetic fibers and innervates the sphincter muscle of pupil as well as the ciliary muscle.
It receives its afferents from the pretectal area of the ipsi- and contralateral side. Its efferents

course in the ipsilateral oculomotor nerve to the second neuron in the ciliary ganglion.

Accessory optic system

Nerves 1

TA Latin: *Nuclei accessorii tractus optici*
TA English: *Accessory nuclei of optic tract*

The accessory optic system (AOS) comprises three terminal nuclei: lateral terminal nucleus, dorsal terminal nucleus and medial terminal nucleus.

It receives its afferents from the accessory optic tract and projects to the dorsal aspect of the inferior olive, which in turn sends afferents to the flocculonodular lobe. In this manner the nucleus plays a role in the coupling of visual information and head movement.

Accessory optic tract

Nerves 1

Subsumed under the expression accessory optic tract are all fibers branching off from the optic tract, medial root, and passing to the three mesencephalic nuclei: dorsal terminal nucleus, lateral terminal nucleus and medial terminal nucleus. These nuclear regions have close contacts with the optic tract nucleus and play a role in smooth pursuit of the eyes as well as in visual-vestibular interactions.

Accessory paresis

Dysfunction of the accessory nerve (XI) results in accessory paralysis rendering it more difficult to lift the arm above shoulder level (trapezius muscle), and turning the head to the unimpaired side is possible only after having successfully contended with resistance (sternocleidomastoid muscle).

Accumbens nucleus

Telencephalon 3

TA Latin: *Nucl. accumbens*
TA English: *Nucl. accumbens*

At the site where the corpus striatum borders on the septal nuclei is situated the accumbens nucleus (septal), which has a structure similar to the corpus striatum, but has unusually intensive fiber connections to the limbic system and hence is viewed as being a link in emotion/motivation and movement.

Adenohypophysis

Diencephalon 3

TA Latin: *Adenohypophysis*
TA English: *Adenohypophysis*
→ Anterior lobe of the hypophysis

ADH (adiuretin, vasopressin)

Diencephalon
→ Supraoptic nucleus

Adrenergic cell group C3 1

A small group beside the adrenergic cell group C2. Is primarily of local significance.

Adrenergic cell groups C1, C2 1

TA Latin: *Cellulae adrenergicae areae postremae et nuclei reticularis ant. (C1, C2)*
TA English: *Adrenergic cells in area postrema and anterior reticular nucleus (C1, C2)*

With the transmitter adrenaline, these cells belong to the monoaminergic cell groups and are part of the lateral reticular formation. C1 and C2 are situated in the immediate vicinity of the noradrenergic cell groups A1 and A2. Efferents to the thalamus, induseum griseum, locus coeruleus and spinal cord where they terminate in autonomic, preganglionic neurons in the interomediolateral nucleus. They are involved in regulation of sympathetic tone.

Adrenergic cell groups C1–C3

Myelencephalon 2

Adrenaline is not a common transmitter in the brain. The three cell groups belong to the monoaminergic cell groups and are situated in the myelencephalon, adrenergic cell groups C1 and C2 in the immediate vicinity of the noradrenergic cell groups A1 and A2.

Efferents go to the hypothalamus, locus coeruleus and spinal cord where they terminate in autonomic, preganglionic neurons in the interomediolateral nucleus.

Agnosia, tactile

→ Inferior parietal lobule

Agraphia

Damage to the angular gyrus causes alexia and agraphia. Those affected are often not able to assign objects they see to specified expressions. Instead of using the correct expression, they engage in circumlocutions in which the expression does not occur.

Akinesia

By generating a facilitating effect on motor information, dysfunction of the globus pallidus results in hypokinesis and poor timing of movements, thus producing motor clumsiness.

Alar lamina

Myelencephalon   1

The alar lamina includes the longitudinal zones SSA, GSA, SVA and GVA, hence all afferent systems.

The basal lamina, conversely, comprises the longitudinal zones GSE, SVE and GVE, hence the efferent nuclei.

Alcoholic encephalopathy

Damage to the mammillary body, e.g. in the case of alcoholic encephalopathy, results in affective impairments and marked loss of perceptivity.

Alexia

Damage to the angular gyrus causes alexia and agraphia. Those affected are often not able to assign objects they see to specified expressions. Instead of using the correct expression, they engage in circumlocutions in which the expression does not occur.

Allocortex

General CNS   2

TA Latin: *Allocortex*

TA English: *Allocortex*

The cerebral cortex consists of two types of tissue:

– allocortex: 3- to 5-layer tissue without a uniform pattern. Frequently encountered in the olfactory system and in the limbic system.
– cerebral cortex: 6-layer nervous tissue accounting for around 95% of the cortex and endowed with a thickness of between 2 mm (visual system) and 5 mm (motor cortex).

Alveus of hippocampus

Telencephalon   2

TA Latin: *Alveus hippocampi*

TA English: *Alveus of hippocampus*

The efferent fibers of the large pyramidal cells of the hippocampus course on the ventricular surface of the hippocampus. This lamella is called the alveus of hippocampus. It bundles to form the fimbria of the hippocampus and later enters the crus of fornix via which the fibers pass as part of the fornix into the direction of the diencephalon (hypothalamus).

Alzheimer's disease

In the presence of presenile or senile demence, especially in the case of Alzheimer's disease, there is marked degeneration of the cells of the basal nucleus Meynert (Ch.4), with concomitant impairment of memory, disorientation, motor unrest and impaired speech.

Ambiens gyrus

Telencephalon   2

The ambiens gyrus borders on the uncus and is partially surrounded by the semilunar gyrus. Both are components of the hippocampus.

Amiculum of olive

Myelencephalon   2

TA Latin: *Amiculum olivare*

TA English: *Amiculum of olive*

Shortly before reaching the dentate nucleus, some afferents of this nucleus form, around the inferior olive, a dense, superficial fiber bundle at the rostral end of the laterodorsal myelencephalon, which is called the amiculum of olive.

Ammon's horn

Telencephalon   3

TA Latin: *Hippocampus proprius*

TA English: *Hippocampus proper*

Ammon's horn together with the dentate fascia and subiculum form the hippocampus, retrocommissural part. It is composed of three-layer allocortex and forms 3 typical structures: digitationes hippocampi, intralimbic gyrus and uncinate gyrus.

The efferents of the large pyramidal cells pass through the alveus hippocampi, fimbria of the hippocampus, crus of fornix and the fornix to the hypothalamus, septum and thalamus.

Amygdala

Telencephalon   3

TA Latin: *Corpus amygdaloideum*

TA English: *Amygdaloid body*
→ Amygdaloid body

Amygdalofugal fibers
Pathways 1
Fiber strands coursing away from the amygdaloid body.

Amygdaloid body
Telencephalon 3
TA Latin: *Corpus amygdaloideum*
TA English: *Amygdaloid body*
The amygdala is a large nuclear complex in the dorsomedial portion of the temporal lobe, at the inferior horn of the lateral ventricle. Reciprocal connections with the rhinencephalon, hypothalamus, thalamus, brainstem and some cortical areas. The amygdaloid body receives highly preprocessed sensory impressions and is responsible for initiation and integration of somatic and autonomic responses, associated with affective behavior.

Amygdaloid body, anterior nucleus
Telencephalon 2
The amygdaloid body is composed of the following nuclear regions:
+ corticomedial group
– anterior nucleus
– cortical nucleus
– medial nucleus
+ basolateral group
– lateral nucleus
– basal nucleus
– basal accessory nucleus
– central nucleus

Amygdalospinal fibers
Pathways 2
Fibers, predominantly from the central amygdaloid nucleus which pass further through the brainstem into the spinal cord, where they may be involved in the regulation of autonomic processes.

Andreas Retzii gyri
Telencephalon 1
Portion of the parahippocampal gyrus situated at the fasciola cinera.

Angular artery
Vessels 2
TA Latin: *A. angularis*
TA English: *Angular artery*
Situated beside the inner angle of the eye and represents the final branch of the facial artery, which arises from the external carotid artery.

Angular gyrus
Telencephalon 2
TA Latin: *Gyrus angularis*
TA English: *Angular gyrus*
The angular gyrus lies in the parietal lobe, and shaped like an angle, it surrounds the posterior end of the superior temporal sulcus
Functionally, it is between the secondary auditory cortex and the area 18. And indeed the angular gyrus does play an important role in linking visual impulses with linguistic concepts.
Damage to the angular gyrus causes alexia and agraphia. Those affected are often not able to assign objects they see to specified expressions. Instead of using the correct expression, they engage in circumlocutions in which the expression does not occur.

Angular vein
Vessels 1
TA Latin: *V. angularis*
TA English: *Angular vein*
A vein situated in the angle of the eye which transports venous blood from the forehead, palpebra superior, cavernous sinus and dorsal nose to the facial vein.

Anosmia
If the olfactory nerve (I) is damaged, e.g. as a sequel of a base of skull injury, depending on the extent of injury, hyposmia or anosmia can ensue. Pungent substances such as ammonia can still be smelt as the nasal mucosa is stimulated, stimulating in turn the trigeminal nerve.

Ansa anastomotica
Vessels 2
The rami cruciantes of the anterior spinal artery run in the shape of a loop around the conus medullaris where they enter the posterior spinal artery.

Ansa lenticularis

TA Latin: *Ansa lenticularis*
TA English: *Ansa lenticularis*
Fiber tract of the subthalamus. The lenticular fasciculus and ansa lenticularis together form the pallidothalamic projection, the biggest efferent of the globus pallidus. The fibers terminate in the ventral lateral thalamic nucleus, which in turn projects to parts of premotor cortex (area 6) and of the supplementary motor area. They arrive at the thalamic nuclei via the thalamic fasciculus.

Ansa peduncularis   2

TA Latin: *Ansa peduncularis*
TA English: *Ansa peduncularis*
In the ansa peduncularis short tracts project from the amygdaloid body to the hypothalamus and to the medial thalamic nucleus.

Ansa peduncularis and ventral amygdalofugal fibers   1

Fibers of the ansa peduncularis and other fibers coming from the amygdaloid body.

Ansoparamedian fissure

Cerebellum   1
TA Latin: *Fissura lunogracilis*
TA English: *Lunogracile fissure*
The ansoparamedian fissure separates from each other the inferior semilunar lobe and gracile lobule.

Anterior amygdaloid nucleus

Telencephalon
TA Latin: *Area amygdaloidea ant.*
TA English: *Anterior amygdaloid area*
→ Amygdaloid body, anterior nucleus

Anterior cerebellar lobe

Cerebellum   2
TA Latin: *Lobus cerebelli ant.*
TA English: *Anterior lobe of cerebellum*
The anterior lobe is the part of the cerebellum rostral to the primary fissure, and is composed of vermis (lingula, central lobule and culmen) as well as hemispheres (quadrangular lobe, anterior part, and ala lobuli centralis). Functionally this subdivision has practically no significance, since the cerebellum evidences a functional arrangement in a vertical direction (vermis, intermediate part, lateral part).

Anterior cerebral artery

Vessels   3
TA Latin: *A. cerebri ant.*
TA English: *Anterior cerebral artery*
Together with the middle cerebral artery, it forms the continuation of the internal carotid artery. Above the optic chiasm, the artery branches off in the direction of the longitudinal fissure of cerebrum. At its entrance, it anastomoses with the anterior cerebral artery of the contralateral side (anterior communicating artery). Hence a distinction is made between the precommunicating part and the postcommunical part. It supplies the entire inside of one hemisphere up to around the occipitotemporal sulcus.

Anterior cerebral artery, postcommunical part

Vessels   3
TA Latin: *A. cerebri ant., pars postcommunicalis*
TA English: *Anterior cerebral artery, postcommunicating part*
The anterior cerebral artery which emerges from the internal carotid artery is divided into a precommunicating part and a postcommunicating part. The latter courses around the genu of the corpus callosum, giving off a series of large vessels before dividing into the pericallosal artery and the callosomarginal artery.
The area supplied is the entire inside of one cerebral hemisphere up to almost the occipitotemporal sulcus.

Anterior cerebral artery, precommunical part

Vessels   3
TA Latin: *A. cerebri ant., pars precommunicalis*
TA English: *Anterior cerebral artery, precommunicating part*
The anterior cerebral artery which emerges from the internal carotid artery is divided into a postcommunicating part and a precommunicating part. The latter gives off branches which supply the optic chiasm and the optic tract. The branching long and short central arteries supply parts of the perforated substance,

the frontal section of the hypothalamus, column of fornix, septum pellucidum, caudate nucleus and internal capsule.

Obstruction of the long central artery results in hemiparesis, paralytic symptoms in the tongue and facial musculature as well as in aphasia.

Anterior cerebral vein

Vessels 3
TA Latin: *V. ant. cerebri*
TA English: *Anterior cerebral vein*

The anterior cerebral vein runs parallel to the anterior cerebral artery. Above the optic chiasm, it branches off in the direction of the longitudinal fissure of the cerebrum. At its entrance, it anastomoses with the contralateral anterior cerebral vein (anterior communicating vein). It drains the inside of one hemisphere to about as far as the central sulcus.

Anterior cerebral veins

Vessels
TA Latin: *Vv. ant. cerebri*
TA English: *Anterior cerebral veins*
→ Anterior cerebral vein

Anterior choroid artery

Vessels 3
TA Latin: *A. choroidea ant.*
TA English: *Anterior choroidal artery*

It belongs to the terminal segments of the internal carotid artery and courses along the optic tract in the direction of the choroid plexus of the lateral ventricle.

The basal part features cortical branches, supplying the parahippocampal gyrus and uncus as well as the dentate gyrus.

The ventricular segment (choroid branch) passes to the choroid plexus of the lateral ventricle and to the choroid plexus of the third ventricle.

Anterior choroid artery, choroid branches (lateral ventricle)

Vessels 2
TA Latin: *A. choroidea ant.,*
Rr. Choroidei ventriculi lat.
TA English: *Anterior choroidal artery,*
choroidal branches to lateral ventricle

Together with the choroid branches of the third ventricle, this branch belongs to the ventricular

segment of the anterior choroid artery.

The branch passes directly to the choroid plexus of the lateral ventricle.

Anterior choroid artery, internal capsular branches

Vessels 1
TA Latin: *A. choroidea ant.,*
Rr. Capsulae internae
TA English: *Anterior choroidal artery,*
branches to internal capsule

On its way to the choroid plexus of the lateral ventricle, the anterior choroid artery sometimes gives off a lateral branch in the direction of the internal capsule.

Anterior column

Medulla spinalis 3
TA Latin: *Funiculus ant.*
TA English: *Anterior funiculus*

The white mater between anterior median fissure and ventral root forms the anterior column, containing:
– anterior pyramidal tract,
– medial longitudinal fasciculus.

Anterior commissure

Telencephalon 3
TA Latin: *Commissura ant.*
TA English: *Anterior commissure*

Fibers from the inferotemporal cortex cross rostral to the splenium and in the anterior commissure to the contralateral side.

Anterior commissure, anterior limb

Diencephalon 2
TA Latin: *Commissura ant., pars ant.*
TA English: *Anterior commissure,*
anterior part

Very narrow branch of the anterior commissure passing to the anterior perforated substance, where it joins the olfactory tract.

Anterior commissure, posterior limb

Diencephalon 2
TA Latin: *Commissura ant., pars post.*
TA English: *Anterior commissure,*
Posterior part

Main part of the anterior commissure passing to the frontal portion of the temporal lobe, hippocampus and amygdaloid body.

Anterior communicating artery

Vessels 3

TA Latin: *A. communicans ant.*

TA English: *Anterior communicating artery*

Small connection segment of both anterior cerebral arteries and important component of the arterial circle of Willis.

Anterior corticospinal tract

Pathways

TA Latin: *Tractus corticospinalis ant.*

TA English: *Anterior corticospinal tract*

→ Anterior pyramidal tract

Anterior ethmoidal artery

Vessels 1

TA Latin: *A. ethmoidalis ant.*

TA English: *Anterior ethmoidal artery*

The artery passing through the ethmoid bone and arising from the ophthalmic artery and forming part of the orbitomeningeal anastomosis.

Anterior ethmoidal foramen

Meninges & Cisterns 1

TA Latin: *Foramen ethmoidale ant.*

TA English: *Anterior ethmoidal foramen*

Point of passage for the similarly named vessels and nerves between the ethmoid bone and frontal bone and resting on the medial wall of the orbita.

Anterior external vertebral venous plexus

Vessels 2

TA Latin: *Plexus venosus vertebralis externus ant.*

TA English: *Anterior external vertebral venous plexus*

Greatly anastomosed venous plexus outside the vertebral column. Via the basivertebral veins, they are connected with the internal vertebral venous plexus.

Anterior forceps

Telencephalon 2

TA Latin: *Forceps minor*

TA English: *Minor forceps*

The commissural fibers running in the splenium of the corpus callosum from the occipital lobe embark on a U-shaped course and are shaped like forceps. They are called the posterior forceps. The anterior forceps is formed from similar U-shaped fibers in the frontal lobe.

Anterior funicle nucleus

Myelencephalon 1

The anterior funicle nucleus and lateral funicle nucleus are situated in the lower myelencephalon and are typical relay nuclei with great projections into the cerebellum. Afferents come from somatosensory tracts from the spinal cord as well as from the somatosensory cortex.

Anterior gray commissure

Medulla spinalis 2

TA Latin: *Commissura grisea ant.*

TA English: *Anterior grey commissure*

In the gray commissure, the nuclear regions, more precisely the intermediate substance, of both halves of spinal cord meet each other.

Whereas the anterior gray commissure runs ventrally to the central canal, the posterior gray commissure passes dorsally to the spinal canal.

Anterior hippocampal veins

Vessels 1

The anterior hippocampal veins are lateral branches of the inferior ventricular vein, ensheathing the digitationes hippocampi.

Anterior horn

Medulla spinalis 3

TA Latin: *Cornu ant.*

TA English: *Anterior horn of the spinal cord*

In the anterior horn are situated the large alpha motoneurons which innervate the skeletal muscles. But the gamma motoneurons responsible for innervation of intrafusal fibers are also encountered here. They play an important role in the refinement of muscle-spindle sensitivity.

The anterior horn evidences a somatotopic arrangement and has two zones:

– medial motor cells,

– lateral motor cells.

Anterior horn of the spinal cord

Medulla spinalis

TA Latin: *Cornu ant.*

TA English: *Anterior horn of the spinal cord*

→ Anterior horn

Anterior hypothalamic area

Diencephalon 1

TA Latin: *Area hypothalamica rostralis*

TA English: *Anterior hypothalamic area*

The anterior hypothalamic area is the term used to designate the anterior hypothalamic nucleus and the surrounding gray mater.

Anterior hypothalamic nucleus

Diencephalon 2

TA Latin: *Nucl. ant. hypothalami*

TA English: *Anterior nucleus of hypothalamus*

The anterior hypothalamic nucleus has myriad diverse afferents, e.g. from the limbic system, other hypothalamic nuclei and the mesencephalon. Efferents go to the surrounding hypothalamic nuclei, but also to the rhombencephalic nuclear regions.

Functionally, the nucleus is involved in regulation of body temperature, respiration and cardiovascular tasks (context: affective defense behavior).

Anterior inferior cerebellar artery

Vessels 3

TA Latin: *A. inf. ant. cerebelli*

TA English: *Anterior inferior cerebellar artery*

Arises from the basilar artery and courses on the lower margin of the pons in the direction of the cerebellum. It crosses the vestibulocochlear nerve where its gives off the labyrinthine branch and then passes into the horizontal fissure of the cerebellum.

By means of its branches, it supplies the lower parts of the hemispheres and of the middle cerebellar peduncle, the flocculus, as well as the choroid plexus of the fourth ventricle.

Anterior intercavernous sinus

Vessels 3

TA Latin: *Sinus intercavernosus ant.*

TA English: *Anterior intercavernous sinus*

The right and left cavernous sinuses are connected via the short anterior intercavernous sinuses and posterior intercavernous sinus. The sinus ring thus formed is called the circular sinus and it surrounds the hypophysis. It receives its venous incoming blood from the latter, as well as from the sphenoid sinus and diaphragma sellae.

Anterior internal frontal artery

Vessels

TA Latin: *A. callosomarginalis, R. Front. anteromed.*

TA English: *Callosomarginal artery, anteromedial frontal branch*

→ Callosomarginal artery, anteromedial frontal branch

Anterior internal vertebral venous plexus

Vessels 2

TA Latin: *Plexus venosus vertebralis internus ant.*

TA English: *Anterior internal vertebral venous plexus*

Venous plexus in the spinal cord. Lies between the layers of spinal dura mater and receives veins from the spinal cord and the spinal meninges.

Anterior limb of internal capsule

Telencephalon 3

TA Latin: *Capsula interna, crus ant.*

TA English: *Anterior limb of internal capsule*

The internal capsule features the following pathways:

posterior limb of internal capsule:
– pyramidal tract,
– superior thalamic peduncle,
– posterior thalamic peduncle,
– parietopontine tract,
– corticotegmental fibers,

anterior limb of internal capsule:
– frontopontine tract,
– anterior thalamic peduncle.

Anterior lobe of the hypophysis

Diencephalon 3

TA Latin: *Adenohypophysis*

TA English: *Adenohypophysis*

The glandular tissue of the anterior lobes produce gonadotropic hormones that regulate the secretion of peripheral hormone glands (ACTH → NNR hormones, TSH → thoroid gland hormones, inter alia) but also effector hormones that act directly (PRL → mammary gland, FSH → gonads inter alia). Regulation is effected via releasing and release-inhibiting factors secreted by neurons of the hypothalamus (infundibular nucleus) into the portal system of the gland.

Anterior lobe of the hypophysis, distal part

Diencephalon   1

TA Latin: *Adenohypophysis, pars distalis*

TA English: *Adenohypophysis, pars distalis*

→ Anterior lobe of the hypophysis, intermediate part

Anterior lobe of the hypophysis, infundibular part

Diencephalon   1

TA Latin: *Adenohypophysis, pars infundibularis*

TA English: *Adenohypophysis, infundibular part*

→ Anterior lobe of the hypophysis, intermediate part

Anterior lobe of the hypophysis, intermediate part

Diencephalon   1

TA Latin: *Adenohypophysis, pars intermedia*

TA English: *Adenohypophysis, pars intermedia*

The anterior lobe of the hypophysis is composed of well-perfused glandular tissue, which can be subdivided into three regions:

– infundibular part (tuberal part).
– intermediate part: narrow strip of tissue containing colloid cysts, depicting a remnant of the Rathke's pouch.
– distal part: biggest part of the gland in which the posterior lobe of the hypophysis is partially embedded.

Anterior median fissure

Medulla spinalis   3

TA Latin: *Fissura mediana ant.*

TA English: *Anterior median fissure*

Fissure in the ventral side of the spinal cord.

Anterior median pontine vein

Vessels

TA Latin: *V. pontis anteromediana*

TA English: *Anteromedian pontine vein*

→ Anterior pontomesencephalic vein (median branch)

Anterior meningeal artery

Vessels   1

TA Latin: *A. ethmoidalis ant., R. Meningeus ant.*

TA English: *Anterior ethmoidal artery, anterior meningeal branch*

Branch of the anterior ethmoidal artery.

Branches off in the anterior cranial fossa and supplies the dura as well as contiguous parts of the falx cerebri.

Anterior occipital sulcus

Telencephalon   3

Relatively constant continuation of the pre-occipital notch.

The sulcus could be called a "three-lobe corner", since here the occipital lobe, parietal lobe and temporal lobe meet.

Anterior olfactory nucleus

Telencephalon   2

TA Latin: *Nucl. olfactorius ant.*

TA English: *Anterior olfactory nucleus*

Small-celled group lying close to the olfactory tubercule in the olfactory trigone.

Anterior paracentral artery

Vessels

TA Latin: *A. pericallosa, R. precunealis*

TA English: *Pericallosal artery, precuneal branch*

→ Precuneal artery

Anterior parolfactory sulcus

Telencephalon   2

TA Latin: *Sulcus parolfactorius ant.*

TA English: *Anterior parolfactory sulcus*

The subcallosal area is enclosed by the anterior parolfactory sulcus and posterior parolfactory sulcus.

Anterior peduncle of thalamus

Diencephalon   2

Corticothalamic and thalamocortical fibers together form the thalamic peduncles:

– anterior peduncle of thalamus: rostral parts of the cerebral cortex and of cingulum,
– inferior peduncle of thalamus: temporal striate cortex, retrosplenial region
– posterior peduncle of thalamus: occipital lobe without area 17 (striate cortex)
– superior peduncle of thalamus: precentral gyrus, postcentral gyrus, prefrontal area.

Anterior perforated substance

Telencephalon   2

TA Latin: *Subst. perforata ant.*
TA English: *Anterior perforated substance*
The anterior perforated substance has a typical, perforated appearance and lies beneath the putamen and globus pallidus, at the site where the olfactory bulb divides into the medial stria and lateral stria (olfactory trigone).
It passes laterally in the direction of the limen insula and contains various nuclei of the secondary, olfactory area and limbic system.

Anterior pontomesencephalic vein

Vessels 1
TA Latin: *V. pontomesencephalica*
TA English: *Pontomesencephalic vein*
The anterior pontomesencephalic vein runs between the basilar artery and the brain tissue from pons and mesencephalon. It receives venous blood from the brain tissue and carries it into the petrosal vein.

Anterior pretectal nucleus

Mesencephalon 1
TA Latin: *Nucl. pretectalis ant.*
TA English: *Anterior pretectal nucleus*
→ Pretectal area

Anterior pyramidal tract

Nerves 3
In the pyramidal decussation 70–90% of the fibers cross to the contralateral side forming the lateral pyramidal tract.
The remaining 10–30% continue their ipsilateral course and descend in the anterior pyramidal tract crossing, however, on entering the gray mater of the spinal cord and innervating also the motoneurons.
The tract extends only as far as the cervical cord.

Anterior quadrigeminal body

TA Latin: *Colliculus sup.*
TA English: *Superior colliculus*
→ Superior colliculus

Anterior radicular artery

Vessels 2
TA Latin: *A. radicularis ant. (Adamkiewiczi)*
TA English: *Anterior radicular artery (Adamkiewicz´s)*
These are short arterial branches from the vertebral artery, which in the cervical region supply

the spinal ganglia and the ventral and dorsal roots of the spinal nerve.

Anterior radicular vein

Vessels 3
The anterior radicular veins penetrate through the ventral root into the spinal canal and thus create an anastomosis between the hemiazygos vein and the anterior spinal vein, flowing on the spinal cord.

Anterior root of spinal nerve

Nerves
TA Latin: *Radix ant. n. spinalis*
TA English: *Anterior root of spinal nerve*
→ Ventral root of the spinal nerve

Anterior spinal artery

Vessels 3
TA Latin: *A. spinalis ant.*
TA English: *Anterior spinal artery*
The unpaired anterior spinal artery and the paired posterior spinal artery arise from the vertebral artery and supply the spinal cord and spinal meninges.

Anterior spinal vein

Vessels
TA Latin: *V. spinalis ant.*
TA English: *Anterior spinal vein*
→ Spinal veins (anterior, lateral, posterior)

Anterior spinocerebellar tract

Cerebellum 3
TA Latin: *Tractus spinocerebellaris ant.*
TA English: *Anterior spinocerebellar tract*
This tract conducts proprio- and exteroceptive impulses (skin receptors, muscle spindles, tendon spindles) of the spinal cord (lumbar cord) to the cerebellum without synapsing in the lateral column. They are the only afferent fibers to enter the cerebellum via the superior cerebellar peduncle.

Anterior temporal artery

Vessels 2
TA Latin: *A. temporalis ant.*
TA English: *Anterior temporal artery*
Arises from the middle cerebral artery, posterior trunk. The middle cerebral artery emerges from

the internal carotid artery.
At the level of the limen insula, it branches off and supplies the superior temporal gyrus, and to some extent also the medial and inferior temporal gyri.

Anterior thalamic nucleus
Diencephalon   3
TA Latin: *Nuclei ant. thalami*
TA English: *Anterior nuclei of thalamus*
This nucleus of the lateral nuclear group is divided into three parts: anteromedial nucleus, anterodorsal nucleus, anteroventral nucleus.
Afferents come from the mammillary body, lateral nucleus and the mammillary body, medial nucleus
In addition, the nucleus has reciprocal connections with the limbic cortex of the cingulate gyrus, the retrosplenial area and the pre- and parasubiculum.
Functions in this regions are emotion, motivation and short-term memory.

Anterior tuberculum of thalamus
Diencephalon   1
TA Latin: *Tuberculum ant. thalami*
TA English: *Anterior thalamic tubercle*
A tuberculum on the anterior side of the thalamus shows the localization of the anterior thalamic nucleus. This hillock is called the anterior tuberculum of thalamus. The choroid tenia of the lateral ventricle passes on via this tuberculum.

Anterior vein of septum pellucidum
Vessels   1
TA Latin: *V. ant. septi pellucidi*
TA English: *Anterior vein of septum pellucidum*
The anterior vein of septum pellucidum carries venous blood from the frontal lobe, septum pellucidum and the roof of the lateral ventricle into the internal cerebral vein.

Anteroinferior diencephalic branches
Vessels
TA Latin: *Aa. centrales anteromed.*
TA English: *Anteromedial central arteries*
→ Anteromedial central arteries

Anterolateral central arteries
Vessels   2
TA Latin: *Aa. centrales anterolat.*
TA English: *Anterolateral central arteries*
Branch off from the sphenoid part of the middle cerebral artery and supply parts of the anterior commissure, of the putamen, of the internal capsule and corona radiata as well as the body of caudate nucleus and part of the head of caudate nucleus.

Anterolateral column
Medulla spinalis   2
TA Latin: *Tractus anterolat.*
TA English: *Anterolateral tract*
The white mater between the ventral root and dorsal root gives rise to the lateral column, containing:
1) anterolateral column with
– anterolateral fasciculus
– parts of the anterior spinocerebellar tract.
2) posterolateral column with
– posterior spinocerebellar tract
– parts of the anterior spinocerebellar tract
– lateral pyramidal tract.

Anterolateral fasciculus
Medulla spinalis   3
TA Latin: *Lemniscus spinalis*
TA English: *Spinal lemniscus*
Somatotopically organized column of the spinal cord containing somatosensory afferents in the direction of the brain. The following tracts course in this column:
– spinotectal tract
– spinothalamic tract
– spino-anular tract
– spino-olivary tract

Anterolateral thalamostriate arteries, lateroinferior diencephalic branches
Vessels
TA Latin: *Aa. centrales anterolat.*
TA English: *Anterolateral central arteries*
→ Anterolateral central arteries

Anteromedial central arteries
Vessels   1
TA Latin: *Aa. centrales anteromed.*
TA English: *Anteromedial central arteries*

Small lateral branches arising from the middle cerebral artery and supplying the surrounding brain tissue, primarily the thalamus.

Anteromedial frontal branch

Vessels
TA Latin: *R. front. anteromed.*
TA English: *Anteromedial frontal branch*
→ Callosomarginal artery, anteromedial frontal branch

Anteromedial thalamostriate arteries; anteroinferior diencephalic branches

Vessels
TA Latin: *Aa. centrales anteromed.*
TA English: *Anteromedial central arteries*
→ Anteromedial central arteries

AOS

Nerves
→ Accessory optic system

Aphasia

Obstruction of the long central artery results in hemiparesis, paralytic symptoms in the tongue and facial musculature as well as in aphasia.

Aphasia, motor

Damage to the inferior frontal gyrus causes motor aphasia. Comprehension of spoken and written language is preserved, with mistakes occurring only on generating one's own language, whose severity correlates with the extent of damage and can range from impaired word-finding ability through agrammatism to complete loss of language.

Aphasia, sensory

Lesions in Wernicke's arear cause sensory aphasia. Unlike motor aphasia, the patients are not capable of comprehending position, meaning of tones and noises. Neither can language be understood, nor can a motor noise with its potential hazard be detected. Musicality is also adversely affected.

Arachnoid

Meninges & Cisterns   3
TA Latin: *Arachnoidea mater*
TA English: *Arachnoid mater*

The arachnoid lies close to the dura mater and together with the pia mater it forms the leptomeninx. Situated between it and the pia mater is the subarachnoid cavity (subarachnoid space) which is filled with CSF. Interspersed in the latter are the eponymous, fine, spiderweblike connective tissue threads (arachnos = spider) connecting the arachnoid with the pia mater.

Arachnoid granulations (Pacchioni)

Meninges & Cisterns   3
TA Latin: *Granulationes arachnoideae*
TA English: *Arachnoid granulations*

The arachnoid granulations are evaginations of the subarachnoid space through the dura mater into the superior sagittal sinus and its lateral lacunae.

CSF is reabsorbed into the blood system at the arachnoid lacunae.

Thus the flow of CSF terminates here, having begun in the choroid plexus.

Arachnoid villi

Meninges & Cisternsh   1
The arachnoid villi are small fingerlike evaginations in the arachnoid granulations. They provide for enlargement of the surface, thus facilitating reabsorption of CSF by the venous blood.

Arcuate fasciculus

Telencephalon   1
TA Latin: *Fasciculus longitudinalis sup.*
TA English: *Superior longitudinal fasciculus*

With its two branches (anterior brachium and posterior brachium), the superior longitudinal fasciculus es lishes connections between virtually all cortical areas. The part of the fasciculus connecting the motor (Broca's) speech center with the sensory (Wernicke's) speech center is called the arcuate fasciclulus.

Arcuate nuclei   1

TA Latin: *Nuclei arcuati*
TA English: *Arcuate nuclei*

The arcuate nuclei are situated on the ventral surface of the pyramid. From here fibers course as external arcuate fibers via the superior cerebellar peduncle to the cerebellum. Other fibers transverse the brainstem and, as medullary striae on the floor of the fourth ventricle, they reach the inferior cerebellar peduncles, via which they pass into the cerebellum.

Area 2v
Telencephalon   2
Experiments in monkeys suggest that cortical areas 2v and 3a are involved in processing vestibular information, thus acting as starting points for taking cognizance of vestibular stimuli.

Area 3a
Telencephalon
→ Area 2v

Area 4
Telencephalon
= Motor cortex = area 4 = primary somatomotor cortex. The precentral gyrus lies in the frontal lobe, directly on the central sulcus. The pyramid cells are encountered here, providing motor control for the contralateral skeletal muscles. All voluntary movements are conducted via this gyrus. It has strict somatotopic arrangement and passes into the longitudinal fissure of cerebrum.

Area 6
Telencephalon
→ **Premotor cortex (area 6)**

Area 8
Telencephalon
Frontal eye field. Plays an important role in voluntary control of eye movement. Part of the premotor cortex.
→ Premotor cortex (area 8)

Area 17 (striate cortex)
Telencephalon   3
Area 17 (striate cortex). The most researched cortical area. Situated deep in the calcarine sulcus, it evidences a white stripe (band of Gennari) which can be seen with the naked eye and features a typical 6-layer structure (typical characteristic of cerebral cortex). Area 17 features highly selective cells which, based on the principle of property extraction, "fish out" those aspects in whose processing they are specialized. Afferents come via the optic radiation from the

lateral geniculate body (LGB) as well as from various cortical areas (also from area 18). Efferents project to the LGB, thalamic pulvinar, superior colliculus, reticular formation and area 18 and area 19.

Areas 17, 18, 19
Telencephalon   1
Area 17 (striate cortex), area 18 and area 19 together form the visual cortex, i.e. the portion of the cerebral cortex directly involved in the processing of visual information.

Area 18
Telencephalon   2
Area 18 (= secondary visual cortex). Further processing of signals from area 17, thalamic pulvinar as well as almost all areas of the ipsilateral hemisphere. Plays a role in interpretation and categorization of visual impressions. Damage results in visual agnosia.

Area 19
Telencephalon   2
Third visual ring. Surrounds area 18 in a ring-shaped manner and is part of the visual association cortex.

Area 22
Telencephalon   2
Part of the superior temporal gyrus, also called auditory cortex. Tasks include, inter alia, the associative further processing of auditory information.

Area 44
Telencephalon
→ Broca´s speech area

Area postrema
Myelencephalon   3
TA Latin: *Area postrema*
TA English: *Area postrema*
Belongs to circumventricular organs and is located in caudal angle to the ventriculus quartus, where it forms an about 1 mm long, but rather narrow elevation. The postremal area is strongly vascularized, during the second half of one's life degenerated, and rich in cells and fibra with different transmitters. Since blood vessels do not

have a blood-brain-barrier here, substances with compulsory barriers reach the nervous tissue as well, which are analyzed here. The AP is the "vomiting center" of the brain, though is also plays an important part in vegetative processes, such as food intake, drinking and cardiovascular regulation.

If intracranial pressure increases, the ensuing concomitant stimulation of the area postrema can elicit emesis.

To curtail elicitation of emesis, dopamine antagonists, inter alia, are used, to act upon the dopamine receptors of the AP, thus suppressing their activating effect.

Area SII

Telencephalon   3

In the lower portion of the postcentral gyrus is situated a cytoarchitectonically slightly modified zone which reaches as far as the lateral sulcus and features a complete representation of the contralateral body half. This area is called the secondary somatosensory cortex, abbreviated SII

Area subenualis

Telencephalon

TA Latin: *Area subcallosa*

TA English: *Subcallosal area*

→ Subcallosal area

Areas 28,35,36

Telencephalon   1

Cortical visual areas in the medial aspect of the hemisphere and contiguous with the hippocampus.

Arterial circle of Willis (left half)

Vessels   3

TA Latin: *Circulus arteriosus cerebri (Willisii)*

TA English: *Cerebral arterial circle*

In the arterial circle of Willis, the large cerebral arteries interconnect in an anastomotic ring. The ring lies in the basal cistern.

Artery of central sulcus

Vessels   3

TA Latin: *A. sulci centralis*

TA English: *Artery of central sulcus*

→ Central sulcus artery

Artery of postcentral sulcus (anterior parietal artery)

Vessels   2

TA Latin: *A. sulci postcentralis*

TA English: *Artery of postcentral sulcus*

Arises from the middle cerebral artery, parietal trunk. The middle cerebral artery emerges from the internal carotid artery.

It ascends in the central sulcus and supplies the pre- and postcentral gyri as well as parts of the parietal gyri.

Artery of precentral sulcus

Vessels   2

TA Latin: *A. sulci precentralis*

TA English: *Artery of precentral sulcus*

Arises from the middle cerebral artery, frontal trunk. The middle cerebral artery for its part emerges from the internal carotid artery.

It ascends in the precentral sulcus and supplies the anterior portion of the precentral gyrus as well as parts of the medial and inferior frontal gyri.

Artery of tectal lamina

Vessels   2

TA Latin: *A. collicularis*

TA English: *Collicular artery*

→ Quadrigeminal artery

Artery of tectal lamina, collicular artery

Vessels   2

TA Latin: *A. collicularis*

TA English: *Collicular artery*

→ Quadrigeminal artery

Artery of the angular gyrus

Vessels   2

Together with the supramarginal artery, it constitutes the terminal segments of the parietal trunk, which for its part emerges from the insular part (M2 segment) of the middle cerebral artery.

It divides in the deep region of the angular gyrus and supplies Heschl's convolution.

The supramarginal artery supplies the supramarginal gyrus as well as the white mater in the vicinity of the insula, including the optic radiation. Dysfunction of the artery of the angular gyrus produces a combination of aphasia, alexia and hemianopsia.

Dysfunction of the supramarginal artery leads to hypoperfusion of the optic radiation, thus causing hemianopsia.

Ascending cervical artery
Vessels   2
TA Latin: *A. cervicalis ascendens*
TA English: *Ascending cervical artery*
Arises from the thyrocervical trunk and supplies deep cervical muscles, scalene muscle, spinal cord, medullary cistern and vertebral canal.

Ascending conjunctive brachium
Cerebellum   1
→ Superior cerebral pedunculus, ascending branch

Ascending lumbar vein
Vessels   1
TA Latin: *V. lumbalis ascendens*
TA English: *Ascending lumbar vein*
Connects the 1st and 2nd lumbar veins with the hemiazygos vein.

Ascending lumbar vein, dorsal branch
Vessels   1
Dorsal branch of the ascending lumbar vein.

Ascending pharyngeal artery
Vessels   2
TA Latin: *A. pharyngea ascendens*
TA English: *Ascending pharyngeal artery*
Arises from the external carotid artery and anastomoses via the jugular foramen with the posterior meningeal branch.
Supplies pharyngeal muscles, Eustachian tube, tonsils, tympanic cavity and cranial meninges of the posterior cranial fossa.

Ascending sacral vein
Vessels
TA Latin: *V. sacralis lat.*
TA English: *Lateral sacral vein*
→ Lateral sacral vein

Association tracts
General CNS
Commissures are fibers which exchange information between the hemispheres. Association pathways are fiber bundles within a hemisphere, while fibers between cerebral cortex and subcortical centers are called projection pathways.

Astereognosis
→ Superior parietal lobule, area 5

Auditory cortex
Telencephalon
The floor of the lateral sulcus is formed by the upper side of the temporal lobe. This almost flat plane is called the temporal plane. It has characteristic, transverse gyri (transverse temporal gyrus (Heschl) which are called Heschl´s transverse convolutions. In this cortex area (area 41 and 42) terminates the auditory tract, hence the term auditory cortex or primary auditory cortex. This area is tonotopically organized and has large efferents in the surrounding auditory cortex.

Auditory radiation
Telencephalon   3
TA Latin: *Radiatio acustica*
TA English: *Acoustic radiation*
The auditory radiation is the term used to designate the system of fibers connecting the medial geniculate body with the auditory cortex.

Auditory tract
Mesencephalon   3
The central auditory tract consists of the following nuclei and tracts: cochlear nuclei, trapezoid body and dorsal acoustic stria, nucleus of the superior olive, lateral lemniscus, inferior colliculus, brachium of inferior colliculus, lateral geniculate body, acoustic radiation and the primary auditory cortex.

Azygos vein
Vessels   3
TA Latin: *V. azygos*
TA English: *Azygos vein*
Transports venous blood from parts of the posterior abdominal wall, vertebral column, spinal cord and some thoracic organs into the superior vena cava.

B

B1

TA Latin: *Cellulae serotoninergicae nuclei raphes pallidi (B1)*
TA English: *Serotoninergic cells in pallidal raphe nucleus (B1)*
→ Raphe pallidus nucleus (B1)

B2

TA Latin: *Cellulae serotoninergicae nuclei raphes obscuri (B2)*
TA English: *Serotoninergic cells in obscurus raphe nucleus (B2)*
→ Raphe obscurus nucleus (B2)

B3

TA Latin: *Cellulae serotoninergicae nuclei raphes magni (B3)*
TA English: *Serotoninergic cells in magnus raphe nucleus (B3)*
→ Raphe magnus nucleus (B3)

B5

TA Latin: *Cellulae serotoninergicae nuclei raphes pontis (B5)*
TA English: *Serotoninergic cells in pontine raphe nucleus (B5)*
→ Raphe pontine nucleus (B5)

B6

TA Latin: *Cellulae serotoninergicae nuclei raphes mediani (B6)*
TA English: *Serotoninergic cells in median raphe nucleus (B6)*
→ Superior central nucleus (B6 + B8)

B7

TA Latin: *Cellulae serotoninergicae nuclei raphes dorsalis (B7)*
TA English: *Serotoninergic cells in dorsal raphe nucleus (B7)*
→ Dorsal raphe nucleus (B7)

B8

→ Superior central nucleus (B6+B8)

Ballism

→ Subthalamic nucleus

Band of Gennari

Telencephalon   3
TA Latin: *Stria occipitalis (Gennari)*
TA English: *Occipital stripe (Gennari)*
Typical white stripe visible with the naked eye which is the characteristic feature of the area 17 (striate cortex) of the primary visual cortex.

Basal accessory nucleus of amygdala

Telencephalon
→ Amygdaloid body

Basal amygdaloid nucleus

Telencephalon
TA Latin: *Nucl. amygdalae basalis*
TA English: *Basal amygdaloid nucleus*
→ Amygdaloid body

Basal ganglia

Telencephalon   3
TA Latin: *Nuclei basales*
TA English: *Basal nuclei*
Gray matter of the telencephalon which does not belong to the cerebral cortex but rather is situated deep in the brain, around the ventricles and thalamus. These include:
– caudate nucleus,
– putamen,
– globus pallidus (=pallidus)
– subthalamic nucleus,
– claustrum.
All these nuclear areas are more-or-less involved in motor function circuits.

Basal lamina

Myelencephalon   1
TA Latin: *Lamina basalis*
TA English: *Basal lamina*
The alar lamina includes the longitudinal zones SSA, GSA, SVA and GVA, hence all afferent systems.
The basal lamina, conversely, comprises the longitudinal zones GSE, SVE and GVE, hence the efferent nuclei.

Basal nucleus Meynert (Ch.4)

Telencephalon   3

TA Latin: *Nucl. basalis subst. basalis*

TA English: *Basal nucleus of basal substance*

In the basal forebrain, between the septum verum and amygdaloid body directly in the middle of the substantia innominata, are situated four groups (Ch.1–Ch.4) comprised of large cholinergic cells, of which Ch.4 are the most pronounced in humans. Afferents come primarily from the limbic system, the projections pass on to all parts of the cerebral cortex. The nuclear region plays a role in the coupling of complex behavioral modes to basic emotional state (motivation).

In the presence of presenile or senile demence, especially in the case of Alzheimer's disease, there is marked degeneration of the cells of the basal nucleus Meynert (Ch.4), with concomitant impairment of memory, disorientation, motor unrest and impaired speech.

Basal pontine nuclei

Pons

TA Latin: *Nuclei pontis*

TA English: *Pontine nuclei*

→ Pontine nuclei

Basal vein (of Rosenthal)

Vessels   2

TA Latin: *V. basalis (Rosenthali)*

TA English: *Basal vein (of Rosenthal)*

The basal vein arises from the union of the anterior cerebral vein and deep middle cerebral vein, passes along the cerebral peduncles and enters the great cerebral vein rostral to the quadrigeminal plate.

Basilar artery

Vessels   3

TA Latin: *A. basilaris*

TA English: *Basilar artery*

The basilar artery courses along the pons and arises from the junction of the vertebral arteries. At the level of the dorsal root of the trigeminal nerve it divides into the two posterior cerebral arteries.

Basilar plexus

Vessels   2

TA Latin: *Plexus basilaris*

TA English: *Basilar plexus*

Venous plexus on the ventral side of the cerebellar pons. The cavernous sinus is connected to the vertebral venous plexuses via this plexus.

Basilar sulcus of pons

Pons   1

TA Latin: *Sulcus basilaris pontis*

TA English: *Basilar sulcus of pons*

Longitudinal groove running along the pontine midline.

Basivertebral vein

Vessels   2

TA Latin: *V. basivertebralis*

TA English: *Basivertebral vein*

The basivertebral veins carry venous blood from the spongy part of the vertebral body into the internal vertebral venous plexus.

Biventer lobule

Cerebellum   2

TA Latin: *Lobulus biventer*

TA English: *Biventral lobule*

The biventer lobule belongs to the posterior lobe and is part of the cerebellar hemispheres. Apart from the areas in proximity to the vermis (intermediate part), the hemispheres belong to the phylogenetically young neocerebellum and receive their afferents via the mossy fibers of the pontocerebellar tract from the pontine nuclei. All hemisphere segments are hence also assigned to the pontocerebellum.

Blindness, cortical

Damage to the visual cortex of one hemisphere leads to anything from disruption of fields of vision (scotoma), directly correlated with the extent of damage, to homonymous hemianopsia (semi-blindness with disruption of one eye field).

If both visual cortices are affected, cortical blindness results. Eye reflexes such as pupillary reflex are preserved, but the cortex-related accommodation reflex is lost.

Body of caudate nucleus

Telencephalon   2

TA Latin: *Corpus nuclei caudati*

TA English: *Body of caudate nucleus*

Body of caudate nucleus.

Body of fornix

Diencephalon   2
TA Latin: *Corpus fornicis*
TA English: *Body of fornix*
Once the efferent fibers of the hippocampus have formed a bundle in the crus of fornix, they unite with the fibers of the contralateral side, thus forming the body of fornix. The latter runs directly beneath the corpus callosum and at the anterior commissure, it divides into the two columns of fornix, which pass into the hypothalamus.

Body of the 11th thoracic vertebra   1

TA Latin: *Corpus vertebrae thoracalis XI*
TA English: *Body of the 11th thoracic vertebra*
Body of the 11th thoracic vertebra.

"Border cell" of the anterior horn of the spinal cord

Medulla spinalis   2
In the ventral root, predominantly motor fibers synapse (alpha and gamma motoneurons).
Only via the border cells of the anterior horn do afferents run from the periphery to the cortical regions of the cerebellum.

Brachiocephalic vein

Vessels   3
TA Latin: *V. brachiocephalica*
TA English: *Brachiocephalic vein*
The brachiocephalic vein arises from the union of the internal jugular vein and the subclavian vein and enters the superior vena cava.

Brachium of inferior colliculus

Mesencephalon   2
TA Latin: *Brachium colliculi inf.*
TA English: *Brachium of inferior colliculus*
Brachium of inferior colliculus. Connects the inferior colliculus with the medial and lateral geniculate bodies of the diencephalon and is part of the central auditory tract.

Brachium of pons

Pons   2
TA Latin: *Pedunculus cerebellaris med.*
TA English: *Middle cerebellar peduncle*
→ Middle cerebellar peduncle

Brachium of superior colliculus

Mesencephalon   2
TA Latin: *Brachium colliculi sup.*
TA English: *Brachium of superior colliculus*
Brachium of superior colliculus. Situated between the superior colliculus and the lateral geniculate body of the diencephalon. Afferent fibers project through the brachium to the superior colliculus, inter alia from the retina, visual cortex and spinal cord.

Brachium restiforme

Cerebellum
TA Latin: *Pedunculus cerebellaris inf.*
TA English: *Inferior cerebellar peduncle*
→ Inferior cerebellar peduncle

Brainstem

General CNS
TA Latin: *Truncus encephali*
TA English: *Brainstem*
Part of the CNS between Medulla oblongata and Mesencephalon. Contains important nuclei for vegetative regulation, cranial nerve nuclei and the cerebellum with its nuclei.
Some authors also see the diencephalon as a part of the brainstem.

Branch of the internal acoustic meatus

Vessels
TA Latin: *A. labyrinthi*
TA English: *Labyrinthine artery*
→ Labyrinthine artery

Broca´s speech area

Telencephalon
The inferior frontal gyrus comprises the following:
– inferior frontal gyrus, orbital part
– inferior frontal gyrus, triangular part
– inferior frontal gyrus, opercular part
In the areas of the frontal gyrus close to the precentral gyrus is situated the premotor cortex, which plays an important role in planning effector voluntary movements and has close interaction with the cerebellum, thalamic nuclei and basal ganglia.
In the inferior frontal gyrus, opercular part, lies the motor speech center (Broca). Here speech is planned but not executed.

Bulbar medial reticular formation

Myelencephalon   1

Medial reticular formation at the level of the myelencephalon.

Bulbospinal fibers

Myelencephalon   2

TA Latin: *Tractus bulboreticulospinalis*
TA English: *Buboreticulospinal tract*

Fibers coming from the brainstem (often called the bulb) to the spinal cord. These fibers constitute the bulbospinal tract, come from the medial reticular formation (e.g. gigantocellular reticular nucleus) and pass on to the ventromedial area of the spinal intermediate zones accommodating the interneurons, which generate an influence on the motoneurons of the axial and proximal muscles of the extremities.

Bulbospinal tract

Pathways

TA Latin: *Tractus bulboreticulospinalis*
TA English: *Bulboreticulospinal tract*

→ Bulbospinal fibers

Bundle of Schütz

Pathways

TA Latin: *Fasciculus longitudinalis post. (Schütz)*
TA English: *Posterior longitudinal fasciculus (Schütz)*

→ Dorsal longitudinal fasciculus (Schütz)

C

C-fibers of dorsal root of the spinal nerve

Medulla spinalis 2

C and A delta fibers innervate primarily nociceptors but also thermo- and some mechanoreceptors. Conversely, A-beta and A-delta fibers innervate touch receptors of the skin.

All these fiber types enter the spinal cord via the dorsal root.

C1

TA Latin: *Cellulae adrenergicae areae postremae et nuclei reticularis anterioris (C1)*
TA English: *Adrenergic cells in area postrema and anterior reticular nucleus (C1)*
→ Adrenergic cell groups C1, C2

C2

TA Latin: *Cellulae adrenergicae areae postremae et nuclei reticularis anterioris (C2)*
TA English: *Adrenergic cells in area postrema and anterior reticular nucleus (C2)*
→ Adrenergic cell groups C1, C2

C3

→ Adrenergic cell group C3

Calcarine sulcus

Telencephalon 3
TA Latin: *Sulcus calcarinus*
TA English: *Calcarine sulcus*
Typical groove running on the median side of the occipital lobe, often entering the parieto-occipital sulcus.

Area 17, the striate cortex, stretches along this sulcus.

Callosomarginal artery

Vessels 3
TA Latin: *A. callosomarginalis*
TA English: *Callosomarginal artery*
Together with the pericallosal artery, the callosomarginal artery arises from the post-communical part of the anterior cerebral artery, which in turn arises from the cerebral part of the internal carotid artery.

Via three branches, it supplies the cingulate gyrus as well as the posterior segment of the superior frontal gyrus.

Callosomarginal artery, anteromedial frontal branch

Vessels 1
TA Latin: *A. callosomarginalis, R. Front. anteromed.*
TA English: *Callosomarginal artery, anteromedial frontal branch*
Together with the callosomarginal artery, intermediomedial frontal branch and the posteromedial frontal branch, this portion forms the terminal segments of the callosomarginal artery.

Together they supply the cingulate gyrus as well as the posterior segment of the superior frontal gyrus.

Callosomarginal artery, intermediomedial frontal branch

Vessels 1
TA Latin: *A. callosomarginalis, R. Front. intermediomed.*
TA English: *Callosomarginal artery, intermediomedial frontal branch*
Together with the callosomarginal artery, anteromedial frontal branch and the postero-medial frontal branch, this portion forms the terminal segments of the callosomarginal artery.

Together they supply the cingulate gyrus as well as the posterior segment of the superior frontal gyrus.

Callosomarginal artery, posteromedial frontal branch

Vessels 1
TA Latin: *A. callosomarginalis, R. Front. posteromed.*
TA English: *Callosomarginal artery, posteromedial frontal branch*
Together with the callosomarginal artery, anteromedial frontal branch and interomedial frontal branch, this portion forms the terminal sections of the callosomarginal artery.

Together they supply the cingulate gyrus as well

as the posterior segment of the superior frontal gyrus.

Calvaria
Skeleton   1
TA Latin: *Calvaria*
TA English: *Calvaria*
The calvaria comprises two hard bony plates and the soft diploe situated between them.
The external lamina is the term used to designate the external plate of the calvaria, and internal lamina refers to the inner plate.
The porous, spongy structure between the calvariae is called diploe.

Candelabra artery
Vessels   1
TA Latin: *A. prefrontalis*
TA English: *Prefrontal artery*
→ Prefrontal artery (candelabra artery)

Carotid canal (lower opening)
Vessels   3
TA Latin: *Canalis caroticus*
TA English: *Carotid canal (lower opening)*
Point of passage of the internal carotid artery and carotid plexus through the petrous part of the temporal bone.

Cauda equina (filia radicularia)
Medulla spinalis   3
TA Latin: *Cauda equina (filia radicularia)*
TA English: *Cauda equina (filia radicularia)*
The ventral and dorsal spinal nerves of the lumbar and sacral cord course in the shape of a horse's tail, parallel to the filum terminale, through the lumbar and sacral portion of the spinal canal to their respective exit points.

Caudal hypothalamic area
Diencephalon   1
TA Latin: *Area hypothalamica post.*
TA English: *Posterior hypothalamic area*
The mammillary body and posterior hypothalamic nucleus are situated in the caudal portion of the hypothalamus.

Caudal olivary nucleus
TA Latin: *Oliva*
TA English: *Inferior olive*
→ Inferior olive

Caudate nucleus
Telencephalon   3
TA Latin: *Nucl. caudatus*
TA English: *Caudate nucleus*
The caudate nucleus and putamen together form the corpus striatum. Both are derived ontogenetically from the same anlagen, but are separated by incoming fibers from the internal capsule.
The corpus striatum is an important inhibitory component of motor movement programs and has manifold connections with the globus pallidus, substantia nigra and the motor cortex.

Caudatolenticular gray bridges
Telencephalon   2
TA Latin: *Pontes grisei caudatolenticulares*
TA English: *Caudolenticular grey bridges*
Columns of cells spreading out between caudate nucleus and putamen.

Caudatopallidal fibers
Telencephalon   1
Fiber connections between the globus pallidus and caudate nucleus.

Cavernous sinus
Vessels   3
TA Latin: *Sinus cavernosus*
TA English: *Cavernous sinus*
Largest sinus at the base of the brain. Lies above the trigeminal ganglion and receives inflow of blood from a large number of deep cerebral veins. It conducts the collected venous blood via the superior petrosal sinus to the sigmoid sinus, where it continues in the direction of the jugular vein.

Cavity of septum pellucidum
Telencephalon   1
TA Latin: *Cavum septi pellucidi*
TA English: *Cave of septum pellucidum*
A fluid-containing cavity between the two laminae of the septum pellucidum.

Central amygdaloid nucleus
Telencephalon
TA Latin: *Nucl. amygdalae centralis*
TA English: *Central amygdaloid nucleus*
→ Amygdaloid body

Central canal

Meninges & Cisterns 3
TA Latin: *Canalis centralis*
TA English: *Central canal*
The central canal is the part of the ventricle system belonging to the spinal cord. It is filled with CSF, begins at the fourth ventricle at the level of the foramen magnum and passes through the entire spinal cord to the beginning of the filum terminale, where it can have a slight swelling.

Central cerebellar nuclei

Cerebellum 3
TA Latin: *Nuclei cerebelli*
TA English: *Cerebellelar nuclei*
The central cerebellar nuclei are located partly in the vermis cerebelli (fastigial nucleus, emboliform nucleus, globose nucleus) and partly in the medulla of the hemispheres (dentate nucleus). Their afferents have their origin in the Purkinje cells of the cerebellar cortex. The cells of the cerebellar hemisphere, lateral part project to the dentate nucleus, the cerebellar hemisphere, intermediate part to the emboliform nucleus and globose nucleus and the vermis cerebelli to the fastigial nucleus.

Central gray matter

General CNS 3
TA Latin: *Substantia grisea centralis*
TA English: *Periaqueductal gray substance*
The central gray matter, also called periaqueductal gray matter, surrounds the mesencephalic aqueduct in the mesencephalon, passing far into the metencephalon. Hence a distinction is made between:
– Central gray matter of mesencephalon,
– Central gray matter of metencephalon.
The centrally located band of cells is an autonomic integration center, akin to the reticular formation. It receives afferents from virtually all parts of the brain and regulates e.g. coordination of the cranial nerve nuclei (e.g. swallowing). By virtue of the close interaction with the limbic system, the central gray matter is also involved in affective fear and flight reactions as well as in pain suppression.

Central gray substance

TA Latin: *Subst. grisea centralis*
TA English: *Central gray substance*

→ Central gray matter

Central lobule

Cerebellum 2
TA Latin: *Lobulus centralis*
TA English: *Central lobule*
The central lobule forms the ventral, upper segment of the vermis cerebelli and rests on the lingula of cerebellum and hence on the fourth ventricle.
Like the entire vermis cerebelli, the central lobule receives its afferents primarily from the spinal cord. It is part of the spinocerebellum = palaeocerebellum.

Central medulla oblongata nucleus

Myelencephalon 2
TA Latin: *Nucl. reticularis centralis*
TA English: *Central reticular nucleus*
Belongs to the lateral reticular formation, i.e. to the parvocellular longitudinal zone of the RF, extending across the entire myelencephalon. Afferents come from the spinal cord, solitary tract, vestibular nuclei and the spinal nucleus of the trigeminal nerve. Efferents go to the gigantocellular reticular formation, the mesencephalic reticular formation as well as the bulbospinal tract in the intermediate substance of the spinal cord.

Central mesencephalic nucleus (Bechterew)

Mesencephalon 1
TA Latin: *Nucl. vestibularis sup. (Bechterew)*
TA English: *Superior vestibular nucleus (Bechterew)*
→ Superior central nucleus (B6 + B8)

Central nucleus of the inferior colliculus

Mesencephalon 2
TA Latin: *Colliculus inf., Nucl. centralis*
TA English: *Central nucleus of inferior colliculus*
Central nucleus of inferior colliculus. Its efferents pass via the brachium of inferior colliculus to the lateral geniculate body. More efferents project to the superior colliculus as well as to the contralateral counterpart.
The nucleus is an important part of reflex chains controlling eye, head and complete-body movements as a function of auditory stimuli.

Afferents originate in the auditory tract and spinal cord.

Central sulcus

Telencephalon   3
TA Latin: *Sulcus centralis*
TA English: *Central sulcus*
The central sulcus separates the frontal lobe and parietal lobe. Situated here is the precentral gyrus (area 4) (precentral gyrus) as well as the primary somatosensory cortex (postcentral gyrus).

Central sulcus artery

Vessels   3
TA Latin: *A. sulci centralis*
TA English: *Artery of central sulcus*
Arises from the middle cerebral artery, parietal trunk. The middle cerebral artery for its part emerges from the internal carotid artery.
It ascends in the central sulcus and supplies the pre- and postcentral gyri.

Central sulcus of the insula

Telencephalon   2
TA Latin: *Sulcus centralis insulae*
TA English: *Central sulcus of insula*
The central sulcus of the insula stretches obliquely across the insula, subdividing the latter into an anterior (short gyri) and posterior (long gyri) part.

Central tegmental tract

Pathways   2
TA Latin: *Tractus tegmentalis centralis*
TA English: *Central tegmental tract*
The central tegmental tract also known as the large longitudinal catecholaminergic bundles, is the most important terminal segment of the extrapyramidal-motor system. Uniting here are efferents from the corpus striatum, globus pallidus, red nucleus, reticular formation, central gray matter of mesencephalon, pons and myelencephalon. The fibers chiefly terminate in the nucleus of the inferior olive from which a powerful tract passes to the cerebellum (olivocerebellar tract). In this manner, a motor feedback system is created, governing coordination of motor control.

Centre médian (Luys)

Diencephalon
TA Latin: *Nucl. centromedianus*
TA English: *Centromedian nucleus*
→ Centromedian nucleus

Centromedian nucleus

Diencephalon   2
TA Latin: *Nucl. centromedianus*
TA English: *Centromedian nucleus*
The centromedian nucleus belongs to the intralaminar thalamic nuclei and receives its afferents from motor and parietal cortex as well as from the globus pallidus. It projects to the putamen, which in turn projects to the globus pallidus. This functional loop conveys polysensory information to the corpus striatum, which is important for execution of correctly oriented motor responses.

Cerebellar body

Cerebellum   1
TA Latin: *Corpus cerebelli*
TA English: *Body of cerebellum*
The cerebellar body is compared with the cerebellar peduncles. It comprises all lobes and lobules and all parts of the vermis cerebelli.

Cerebellar commissure

Cerebellum   3
TA Latin: *Commissura cerebelli*
TA English: *Cerebellar commissure*
The two cerebellar hemispheres communicate via long commissural fibers. The associated bundle of fibers crossed the vermis cerebelli close to the fastigial nucleus. The preceding part is called the anterior cerebellar commissure and the succeeding part is known as the posterior cerebellar commissure (Stilling).

Cerebellar cortex

Cerebellum   2
TA Latin: *Cortex cerebelli*
TA English: *Cerebellar cortex*
Just like the cerebrum, the cerebellum also evidences a pronounced cortical structure. The gray nuclear cortex is greatly folded and interspersed with white, fiber-containing matter.
The cerebellar cortex has a typical cytoarchitecture whose chief components are

Purkinje cells, granular cells, basket cells and Golgi cells.
The cerebellar cortex compares motor programm with motor action and optimizes the motor programm.

Cerebellar hemisphere

Cerebellum 3
TA Latin: *Hemispherium cerebelli*
TA English: *Hemisphere of cerebellum*
The cerebellum can be divided into three parts:
– hemispheres (cerebellar hemispheres),
– vermis cerebelli,
– peduncles (cerebellar peduncles).
The hemispheres have a pronounced cortical structure (cerebellar cortex) rising like a tree from the central matter (medullary body of cerebellum) and is called arbor vitae, the tree of life.

Cerebellar hemisphere, intermediate part

Cerebellum 3
The regions of the cerebellar hemisphere that are close to the vermis are called the cerebellar hemisphere, intermediate part. This runs around 1 cm to the right and left of the vermis, and like the latter it receives its afferents primarily from the spinal cord (spinocerebellum).
As opposed to the vermis, the Purkinje cells of the intermediate part project to the interpositus nucleus and not, as in the case of the vermis, to the fastigial nucleus.

Cerebellar hemisphere, lateral part

Cerebellum 3
The cerebellar hemisphere is subdivided into the intermediate part close to the vermis and the remaining lateral part. This has resulted from important functional observation, indicating that the Purkinje cells located in this lateral part have a common projection area, i.e. the dentate nucleus, while conversely the Purkinje fibers of the of the cerebellar hemisphere, intermediate part, project to the interpositus nucleus.

Cerebellar nuclei

Cerebellum 3
TA Latin: *Nuclei cerebelli*
TA English: *Cerebellar nuclei*
Subsumed under this collective term are four central cerebellar nuclei:

– dentate nucleus,
– fastigial nucleus,
– emboliform nucleus,
– globose nucleus.

Cerebellar veins

Eye 2
TA Latin: *Vv. cerebelli*
TA English: *Cerebellar veins*

Cerebellomedullary cistern

Meninges & Cisterns 3
TA Latin: *Cisterna cerebellomedullaris*
TA English: *Cerebellomedullary cistern*
The cerebromedullary cistern is the largest cistern and is formed primarily by the left and right tonsil of cerebellum. It passes along the vermis until the upper side of the cerebellum, where it joins the superior cerebellar cistern. Inferiorly, it joins the subarachnoid spinal cistern. The cistern is connected with the fourth ventricle via the median aperture of the fourth ventricle.
The cerebellomedullary cistern is clinically relevant since it can be punctured (suboccipital).

Cerebellopontine angle cistern

Meninges & Cisterns 2
TA Latin: *Cisterna pontocerebellaris*
TA English: *Pontocerebellar cistern*
→ Pontocerebellar cistern

Cerebellorubral tract

Cerebellum 3
TA Latin: *Pedunculus cerebellaris sup.*
TA English: *Superior cerebellar peduncle*
→ Superior cerebellar peduncle

Cerebellothalamic tract

Cerebellum 3
TA Latin: *Pedunculus cerebellaris sup.*
TA English: *Superior cerebellar peduncle*
→ Superior cerebellar peduncle

Cerebellovestibular fibers

Cerebellum 2
The cerebellovestibular fibers are small fiber bundles that pass directly without synapsing in the fastigial nuclei, from the cortical regions of the flocculus and nodulus via the superior cerebellar peduncle to the vestibular nuclei.

Globally, they are designated as the cerebello-vestibular tract.

Cerebellovestibular tract

Cerebellum   2

The cerebellovestibular tract is the global name for all fibers coming directly, i.e. without synapsing, to the cerebellar nuclei from the cortical region of the vestibulocerebellum (flocculus+nodulus) to the vestibular nuclei. They do so via the inferior cerebellar peduncle and reach all four vestibular nuclei.

These "fibers" are also called the cerebellovestibular fibers.

Cerebellum

Cerebellum   3

TA Latin: *Cerebellum*

TA English: *Cerebellum*

Cerebellum is composed of a centrally situated vermis ("worm") and the two hemispheres. It is responsible above all for planning motor programs and for preserving equilibrium.

Cerebellum, zonal arrangement

Cerebellum   3

Although the cerebellum features horizontal organization by virtue of its fissures, it is divided functionally into three vertical zones: the central part corresponds to the vermis cerebelli and projects to the fastigial nucleus. The intermediate part is a stip of hemisphere that is less than 1 cm wide to the left and right of the vermis. It projects to the interpositus nucleus. The lateral part, the remaining hemisphere region, projects to the dentate nucleus.

Cerebral aqueduct

Meninges & Cisterns   3

TA Latin: *Aquaeductus mesencephali*

TA English: *Aqueduct of midbrain*

→ Mesencephalic aqueduct

Cerebral cortex

Telencephalon   3

TA Latin: *Cortex cerebri*

TA English: *Cerebral cortex*

This is the term used for the entire gray matter contained in the cortical region of the cerebrum. According to cytoarchitecture you differentiate isocortex and allocortex. A more detailed analy-sis reveals nearly 50 different cortical areas, the so called Brodmann Areas. The cerebral cortex ist divided into 5 lobes: frontal, parietal, temporal, occipital and limbic lobe. The cerebral cortex is essential for cognition, memory, speach and voluntary movement.

Cerebral crus

TA Latin: *Crus cerebri*   3

TA English: *Cerebral crus*

→ Cerebral peduncle

Cerebral peduncle

Mesencephalon   3

TA Latin: *Pedunculus cerebri*

TA English: *Cerebral peduncle*

Above the pons are two large, v-shaped parallel fiber bundles, containing efferents descending from the cerebral cortex in the direction of the brainstem and spinal cord. These two strands are called cerebral peduncles.

Cerebral trunk

General CNS   3

Brainstem. Is composed of the three segments myelencephalon, metencephalon (cerebellum + pons) and mesencephalon.

Cerebral veins

Vessels   2

TA Latin: *Vv. encephali*

TA English: *Cerebral veins*

Deep and cortical cerebral veins.

Cerebromeningeal anastosmosis

Vessels   2

Connection between the middle meningeal artery and the callosomarginal artery.

Cerebrospinal fluid

Meninges & Cisterns   3

TA Latin: *Liquor cerebrospinalis*

TA English: *Cerebrospinal fluid*

Cerebrospinal fluid (CSF). A distinction is made between two systems: CSF is produced in the ventricles (choroid plexus) and, under control at the apertures (median and lateral) of the fourth ventricle, it escapes into the subarachnoid space, where it is reabsorbed via the arachnoid granulations (Pachioni) into the venous blood system. This subarachnoid space can be punctured at various sites. CSF is colorless, low in protein and almost free of cells.

Cerebrum (outer surface)

Telencephalon   3

TA Latin: *Cerebrum*

TA English: *Cerebrum (external features)*

At a deep level, is composed of the basal ganglia and peripherally of the greatly folded cerebral cortex, which is subdivided into two hemispheres.

Here all "higher" brain functions such as voluntary motor control, motor and sensory speech, cognition, visual and auditory system, superficial and deep sensibility are processed.

Cervical enlargement

Medulla spinalis   2

TA Latin: *Intumescentia cervicalis*

TA English: *Cervical enlargement*

The spinal cord evidences two enlargements: the cervical enlargement in the cervical region and the lumbosacral enlargement in the lumbar region. The fibers of the upper and lower extremities synapse in the enlargements.

Ch.12

TA Latin: *Cellulae cholinergicae nuclei septi medialis (Ch1)*

TA English: *Cholinergic cells of medial septal nuclei (Ch1)*

→ Medial septal nucleus (Ch1)

Chiasmatic cistern

Meninges & Cisterns   3

TA Latin: *Cisterna chiasmatica*

TA English: *Chiasmatic cistern*

The chiastmatic cistern surrounds the optic chiasm and infundibulum and passes along the optic nerve into the orbita.

Cholinergic cell groups Ch1–Ch6   1

TA Latin: *Cellulae cholinergicae Ch1–Ch6*

TA English: *Cholinergic cells Ch1–Ch6*

The cell groups, which use acetylcholine as transmitter, are compared with the monoaminergic cell groups and contain four groups (Ch.1 – Ch.4) in the basomedial telencephalon and two groups (Ch.5 + Ch.6) in the brainstem.

Chorea

Damage to the corpus striatum results in the typically manifest symptoms of chorea, due to disinhibition of the globus pallidus and substantia nigra. Chorea is characterized at an advanced stage by hyperkinesia especially of the distal extremities' musculature and of the face. Dystonic syndrome (e.g. retrocollis, spastic torticollis) or athetosis are also encountered.

Choroid branches of the fourth ventricle (of posterior inferior cerebellar artery)

Vessels

TA Latin: *Rr. choroidei ventriculi quarti a. Inf. post. cerebelli*

TA English: *Choroidal branches to fourth ventricle of posterior inferior cerebellar artery*

→ Posterior inferior cerebellar artery, choroid branch of fourth ventricle

Choroid fissure

Meninges & Cisterns   1

TA Latin: *Fissura choroidea*

TA English: *Choroidal fissure*

If one removes the choroid plexus, the teniae are clearly visible as a torn edge. The cleft between these edges is called the choroid fissure.

Choroid line

Meninges & Cisterns   2

TA Latin: *Taenia choroidea*

TA English: *Choroid line*

The choroid plexus of the lateral ventricle is secured to the thalamic complex by means of the choroid tenia. At the third ventricle, it joins the tenia of the thalamus.

Parallel to it stretches the tenia of the fornix. In between, is a membrane (choroid tela of the lateral ventricle) supporting the choroid plexus of the lateral ventricle.

Choroid plexus of the fourth ventricle

Meninges & Cisterns   3

TA Latin: *Plexus choroideus ventriculi quarti*

TA English: *Choroid plexus of fourth ventricle*

→ Choroid plexus of the lateral ventricle

Choroid plexus of the lateral ventricle

Meninges & Cisterns   3

TA Latin: *Plexus choroideus ventriculi lat.*

TA English: *Choroid plexus of lateral ventricle*

The choroid plexus produces CSF.
It is composed of a single layer of lamina with villi, the choroid epithelial lamina, in which the CSF is also produced, as well as the dorsal vascularized choroid tela.

Choroid plexus of the third ventricle
Meninges & Cisterns   3
TA Latin: *Plexus choroideus ventriculi tertii*
TA English: *Choroid plexus of third ventricle*
→ Choroid plexus of the lateral ventricle

Choroid tela (all ventricles)
Meninges & Cisterns   2
TA Latin: *Tela choroidea (aller Ventrikel)*
TA English: *Tela choroidea (all ventricles)*
External, vascularized layer of connective tissue of a choroid plexus. On its inside is a single layer sheet of choroid epithelium, forming villi and responsible for CSF production.

Ciliar body
Eye   2
TA Latin: *Corpus ciliare*
TA English: *Ciliary body*
Ciliary body of the eye. Contraction of the circular ciliary muscle results in relaxation of the lens ligament (zonal fibers), so that the lens can follow its inner elasticity and thicken. This increases its refractive power, needed for focusing on close objects. If conversely, the ciliary muscle is relaxed, the eye is distant accommodated.

Ciliary ganglion
Nerves   3
TA Latin: *Ganglion ciliare*
TA English: *Ciliary ganglion*
Parasympathetic ganglion, some 2 cm behind the eyeball. The postganglionic fibers innervate, inter alia, two intraocular muscles:
– ciliary muscle (accommodation),
– sphincter of pupil muscle (adaptation).

Ciliospinal center
Medulla spinalis   2
Pupil dilatation which is controlled by the sympathetic nervous system has its origin in the thoracic cord, in the ciliospinal center. The afferents of this center come from an unknown

source in the brainstem, while efferents go to the superior cervical ganglion where they synapse and project as far as the pupil.

Cingular branch
Vessels
TA Latin: *R. cingularis*
TA English: *Cingular branch*
→ Callosomarginal artery

Cingulate gyrus
Telencephalon   3
TA Latin: *Gyrus cinguli*
TA English: *Cingulate gyrus*
An important component of the limbic system. Comprises the subcallosal area, the actual cingulate area and the retrosplenial cortex.
Has connections with the hypothalamus, corpus striatum and association cortex.
Plays a role in behavioral modes for food intake, drive and motivation.
Impairment of the cingulate gyrus, e.g. due to a tumor deep in the longitudinal fissure of cerebrum, causes pronounced sedation of the patient, loss of psychic and motor initiative as well as indifference and psychic apathy.

Cingulate sulcus
Telencephalon   3
TA Latin: *Sulcus cinguli*
TA English: *Cingulate sulcus*
A sulcus visible in median section, which surrounds the cingulate gyrus and thus encloses the limbic lobe. In the transitional region between occipital lobe and parietal lobe it joins the marginal part and ascends to the margin of the hemisphere.

Cingulate sulcus, marginal part
Telencephalon   1
TA Latin: *Sulcus cinguli, R. marginalis*
TA English: *Cingulate sulcus, marginal branch*
A lateral branch of the cingulate sulcus ascending to the margin of the hemisphere.

Cingulomarginal artery
Vessels
TA Latin: *A. callosomarginalis*
TA English: *Callosomarginal artery*
→ Callosomarginal artery

Cingulomarginal artery, cingular branch

Vessels
TA Latin: *A. callosomarginalis*
TA English: *Callosomarginal artery*
→ Callosomarginal artery

Cingulothalamic artery

Vessels   2
The cingulothalamic artery branches from the middle occipital artery and passes into the space between splenium of the corpus callosum and pineal body. Here it gives off a superior thalamic branch as well as a dorsal branch of the corpus callosum. The latter courses largely parallel to the cingulum and anastomoses with the pericallosal artery, posterior branch, thus completely surrounding the corpus callosum.

Cingulothalamic artery, branch of the dorsal corpus callosom

Vessels   2
TA Latin: *A. occipitalis med. , R. corporis callosi dors.*
TA English: *Medial occipital artery, dorsal branch to corpus callosom*
Lateral branch of the cingulothalamic artery which courses parallel to the cingulum and anastomoses with the pericallosal artery, posterior branch.

Cingulothalamic artery, superior thalamic branch

Vessels   1
The superior thalamic branch arises from the cingulothalamic artery, which is a side projection of the middle occipital artery. In conjunction with the lateral occipital artery, this in turn forms the terminal part (P4) of the posterior cerebral artery.
The thalamic branch penetrates the upper thalamic tissue, thus supplying it.

Cingulum

Telencephalon   2
TA Latin: *Cingulum*
TA English: *Cingulum*
The cingulum is a strong bundle of association pathways of varying length that connects different cortical centers of a hemisphere. It is situated on the lower margin of the cingulate gyrus.

Circadian rhythm

→ Suprachiasmatic nucleus

Circular sulcus of the insula

Telencephalon   3
TA Latin: *Sulcus circularis insulae*
TA English: *Circular sulcus of insula*
The circular sulcus of the insula stretches as suggested by the name in a circle around the insula, demarcating it from the surrounding lobes.

Circumventricular organs

3
TA English: *Circumventricular organs*
An area situated in the ventricle wall in which the blood-brain barrier is restricted and the ependyma is specialized in performing certain tasks. These areas include:
– posterior lobe of the hypophysis,
– area postrema,
– organum vasculosum of the lamina terminalis,
– subfornical organ,
– pineal body.

Cistern of central sulcus

Meninges & Cisterns   2
The cistern of central sulcus is formed by the pronounced central sulcus, which separates the frontal lobe and parietal lobe.

Cistern of corpus callosum

Meninges & Cisterns   1
TA Latin: *Cisterna pericallosa*
TA English: *Pericallosal cistern*
→ Pericallosal cistern

Cistern of lamina terminalis

Meninges & Cisterns   1
TA Latin: *Cisterna laminae terminalis*
TA English: *Cistern of lamina terminalis*
The cistern of the lamina terminalis surrounds the space between the two paraterminal gyri. The anterior parolfactory sulcus and the cingulate gyrus are contiguous with this cistern.

Cistern of tectal lamina

Meninges & Cisterns   1
TA Latin: *Cisterna quadrigeminalis*
TA English: *Quadrigeminal cistern*
The cistern of tectal lamina is a cistern immediately before the quadrigeminal lamina (=tectum of mesencephalon). Superiorly, it is bordered by

the pineal body and, posteriorly, by the anterior cerebellar lobe.

Coursing with it is the basal vein (of Rosenthal).

Cistern of the great cerebral vein

Meninges & Cisterns 1

TA Latin: *Cisterna venae magnae cerebri*

TA English: *Cistern of the great cerebral vein*

Cistern around the great cerebral vein (Galen), which here receives the internal cerebral vein and basal vein and joins the straight sinus.

Cistern of the internal acoustic meatus

Meninges & Cisterns 1

The external acoustic pore (porion), upper edge. Around the facial nerve (VII) and cochlear nerve (VIII) is also formed a cistern, the cistern of the internal acoustic meatus.

Cistern of the lateral fossa of the cerebrum

Meninges & Cisterns 2

TA Latin: *Cisterna fossae lat. cerebri*

TA English: *Cistern of lateral cerebral fossa*

Cistern formed by the large lateral sulcus, embedded deep in the insula.

Cistern of the medulla oblongata and spine

Meninges & Cisterns

→ Medullary cistern

Cistern of the sylvian fossa

Meninges & Cisterns

TA Latin: *Cisterna fossae lat. cerebri*

TA English: *Cistern of lateral cerebral fossa*

→ Cistern of the lateral fossa of the cerebrum

Cistern of the transverse fissure

Meninges & Cisterns 1

The cistern of the transverse fissure has its opening between the pineal body and splenium of the corpus callosum and passes along the crus of fornix deep into the diencephalon.

The internal cerebral vein courses with it.

Cisterna ambiens

Meninges & Cisterns 2

TA Latin: *Cisterna ambiens*

TA English: *Cisterna ambiens*

Conspicuous extensions of the space between arachnoid and pia mater are called cisterns.

The cisterna ambiens surrounds the pineal body and the quadrigeminal lamina.

Cisterna magna

Meninges & Cisterns 3

TA Latin: *Cisterna cerebellomedullaris post.*

TA English: *Posterior cerebellomedullary cistern*

→ Cerebellomedullary cistern

Cisterna valleculae cerebri

Meninges & Cisterns 1

Vallis=valley.

The cisterna valleculae cerebri spreads across the valley that is situated deep in the temporal pole, around the limen insula.

Cisterna veli interpositi

Meninges & Cisterns 1

→ Cistern of the transverse fissure

Cisterns

Meninges & Cisterns 3

TA Latin: *Zisternen*

TA English: *Cisterns*

Extraneural liquor space.

Clarke's column (Stilling-Clarke)

Medulla spinalis 2

TA Latin: *Nucl. thoracicus post. (Stilling-Clarke)*

TA English: *Posterior thoracic nucleus (Stilling-Clarke)*

→ Thoracic nucleus

Claustrum

Telencephalon 2

TA Latin: *Claustrum*

TA English: *Claustrum*

The claustrum, which is a lamina of gray matter running parallel to the putamen and stretching as far as the amygdaloid body, is situated between the external capsule and the extreme capsule. The claustrum is of a pronounced heterogeneous expression, but with reciprocal connections with contra- and ipsilateral cortical areas being encountered. Therefore, and due to the multimodal afferents, associative functions are attributed to the claustrum.

Coccygeal segment of the spinal cord

Medulla spinalis   1
TA Latin: *Pars coccygea medullae spinalis*
TA English: *Coccygeal part of spinal cord*
Coccygeal cord. The part of the spinal cord surrounding the coccyx.

Cochlear nerve

Nerves   3
TA Latin: *N. cochlearis*
TA English: *Cochlear nerve*
First section of the auditory tract. Is part of vestibulocochlear nerve (VIII) and goes from the spiral ganglion (1st neuron of the auditory tract) to the cochlear nuclei. The fibers are organized in strict tonotopic fashion (acc. to tone frequencies).

Cochlear nuclei

Pons   3
TA Latin: *Nuclei cochleares*
TA English: *Cochlear nuclei*
Having entered the brain tissue, the cochlear nerve branches into two parts: fibers from the lower segment of the cochlea course dorsally to the dorsal cochlear nucleus, the fibers from the upper cochlear region pass ventrally to the ventral cochlear nucleus.
The nuclei have strict tonotopic organization. Efferents go either directly to the contralateral inferior colliculus or, after synapsing, to the nuclei of the trapezoid body, nucleus of the superior lateral olive, medial nucleus of the superior olive, nucleus of the trapezoid body and lateral lemniscal nuclei.

Coeruleospinal tract

Mesencephalon   2
TA Latin: *Tractus caeruleospinalis*
TA English: *Caeruleospinal tract*
In the dorsal noradrenergic bundle of the locus coeruleus, fibers run in the direction of the spinal cord where they run in the lateral column and pass to all segments of the spinal cord, terminating in the posterior horn, in the anterior horn and in the intermediate substance.
This portion of the coerulean efferents are globally called the coerulospinal tract.

Collateral circulation, arterial

Vessels   1

Collaterals are vessels aspiring to supply a region similar to that of the main vessel and which in the event of obstruction of the main vessel, can act as a substitute blood stream.

Collateral circulation, venous

Vessels
→ Collateral circulation, arterial

Collateral sulcus

Telencephalon   2
TA Latin: *Sulcus collateralis*
TA English: *Collateral sulcus*
Sulcus between the parahippocampus gyrus and the lateral occipitotemporal gyrus.
The perirhinal cortex is situated deep in the sulcus on the side of the parahippocampal gyrus.

Collicular artery

Vessels
TA Latin: *A. collicularis*
TA English: *Collicular artery*
→ Quadrigeminal artery

Column of fornix

Diencephalon   2
TA Latin: *Columna fornicis*
TA English: *Column of fornix*
At the anterior commissure, the body of fornix divides into the two columns of fornix which both pass into their respective hypothalamus.

Commissure

General CNS   3
TA Latin: *Commissura*
TA English: *Commissure*
Commissure interconnects brain centers of both hemispheres.
Commissures are fibers which exchange information between the hemispheres. Association pathways are fiber bundles within a hemisphere, while fibers between cerebral cortex and subcortical centers are called projection pathways.

Commissure of the fornix

Diencephalon   2
TA Latin: *Commissura fornicis*
TA English: *Commissure of the fornix*
In the crus of fornix a few hippocampal efferents pass to the contralateral side. These fibers form

the commissure of the fornix which passes directly in the inner curvature of the splenium of the corpus callosum.

Commissure of the inferior colliculus

Mesencephalon 2

TA Latin: *Commissura colliculi inf.*

TA English: *Commissure of inferior colliculus*

Unites both inferior colliculi. Fibers of the lateral lemniscus also pass to the contralateral side here.

Commissure of the superior colliculus

Mesencephalon 1

TA Latin: *Commissura colliculi sup.*

TA English: *Commissure of superior colliculus*

Reciprocal connections between both superior colliculi.

Common carotid artery

Vessels 3

TA Latin: *A. carotis communis*

TA English: *Common carotid artery*

The common carotid artery emerges directly from the aortic arch and, at the level of the third cervical vertebra, it divides into the larger internal carotid artery and the smaller external carotid artery. The former provides primarily for intracranial supply, and the latter for extracranial supply.

Common iliac vein

Vessels 3

TA Latin: *V. iliaca communis*

TA English: *Common iliac vein*

Arises from the junction of external iliac vein and internal iliac vein and flows into the inferior vena cava.

Communicating branch (spinal nerve)

Nerves 2

TA Latin: *R. communicans albus nervorum spinalium*

TA English: *White ramus communicans*

Sympathetic fibers emerge from the spinal nerves after passing through the intervertebral foramen and then course through the white communicating ramus to the sympathetic ganglion, where they either synapse to return via the gray communicating ramus, or continue further, synapsing in proximity to an organ.

Communicating branch to the meningeal branch

Meninges & Cisterns 1

TA Latin: *R. communicans cum ramo meningeo*

TA English: *Communicating branch with meningeal branch*

Lateral branch via which the lateral meningeal branches are interconnected.

Confluence of the sinuses

Vessels 3

TA Latin: *Confluens sinuum*

TA English: *Confluence of sinuses*

As indicated by the name, various sinuses unite in the occipitally located confluence of the sinuses, including the superior sagittal sinus, straight sinus and tentorial sinus.

The two transverse sinuses serve as a drain, transporting venous blood in the direction of the superior jugular vein, which it joins at the superior bulb of the jugular vein.

Conjunctive brachium

Cerebellum 1

TA Latin: *Pedunculus cerebellaris sup.*

TA English: *Superior cerebellar peduncle*

→ **Superior cerebellar peduncle**

Connecting vein

Vessels 1

Connects the superior choroid vein and superior thalamic vein.

Conus medullaris

Medulla spinalis 3

TA Latin: *Conus medullaris*

TA English: *Conus medullaris*

At the level of the first and second lumbar vertebrae, the spinal cord unites in a cone-shaped manner. This is called the conus medullaris. The cone enters the filum terminale.

Corona radiata

Telencephalon 3

TA Latin: *Corona radiata*

TA English: *Corona radiata*

The fibers from the internal capsule are crossed by the fibers of the corpus callosum, causing them to deviate in a fan-shaped manner.

Corpus callosum

Telencephalon   3
TA Latin: *Corpus callosum*
TA English: *Corpus callosum*
The largest commissure of the brain. Connects the two halves of the cerebrum and forms the floor of the longitudinal fissure of cerebrum. Consists of four parts: splenium, trunk, genu and rostrum.

Corpus striatum

Telencephalon   3
TA Latin: *Corpus striatum*
TA English: *Corpus striatum*
The corpus striatum is divided into the caudate nucleus and putamen. In its capacity of central inhibitory relay station for motor signals, it receives activating afferents from the cerebral cortex and centromedian nucleus, and inhibitory signals from the substantia nigra. Inhibitory GABAergic efferents go to the globus pallidus and the substantia nigra. The main function is predominantly inhibitory processing of cortical motor signals.

Damage to the corpus striatum results in the typically manifest symptoms of chorea, due to disinhibition of the globus pallidus and substantia nigra. Chorea is characterized at an advanced stage by hyperkinesia especially of the distal extremities' musculature and of the face. Dystonic syndrome (e.g. retrocollis, spastic torticollis) or athetosis are also encountered.

Cortical amygdaloid nucleus

Telencephalon
TA Latin: *Nucl. amygdalae corticalis*
TA English: *Cortical amygdaloid nucleus*
→ Amygdaloid body

Corticobulbar tract

Pathways   2
Fibers coursing from cerebral cortex to brainstem (also called bulb).

Corticonuclear fibers

Pathways   1
TA Latin: *Fibrae corticonucleares*
TA English: *Corticonuclear fibres*
On reaching the vicinity of their target region, the fibers of the pyramidal tract disengage themselves from the tract and form individual fibers, which are called corticonuclear fibers.

Corticonuclear tract

Pathways   3
TA Latin: *Fibrae corticonucleares*
TA English: *Corticonuclear fibres*
→ Corticonuclear fibers

Corticospinal tract

Pathways   3
TA Latin: *Tractus corticospinalis*
TA English: *Corticospinal tract*
→ Pyramidal tract

Corticostriatal projection

Telencephalon   1
Direct and indirect projections from the cerebral cortex to the corpus striatum.

Corticotegmental fibers

Pathways   1
Cortical fibers from the contralateral motor cortex, projecting to the lateral reticular formation and, together with the projections from the contralateral red nucleus, are involved in a system regulating fine adjustment and alignment of effector motor responses.

Corticovestibular tract

Pathways   1
Cortical projections to the vestibular nuclei.

Costocervical trunk

Vessels   2
TA Latin: *Truncus costocervicalis*
TA English: *Costocervical trunk*
Arises from the subclavian artery and joins the:
– deep cervical artery,
– superior intercostal artery

Costotransverse foramen

Meninges & Cisterns   1
TA Latin: *Foramen costotransversarium*
TA English: *Costotransverse foramen*
The costotransverse foramen is bordered inferiorly by the upper side of the succeeding rib, superiorly, by the transverse process, inwards by the body of vertebra and outwards by the costotransverse ligament.

Cranial nerves

Nerves   1
TA Latin: *Nn. craniales*
TA English: *Cranial nerves*
Cranial nerves:
– Olfactory nerve (I)
– Optic nerve (II)
– Oculomotor nerve (III)
– Trochlear nerve (IV)
– Trigeminal nerve (V)
– Abducens nerve (VI)
– Facial nerve (VII)
– Vestibulocochlear nerve (VIII)
– Glossopharyngeal nerve (IX)
– Vagus nerve (X)
– Accessory nerve (XI)
– Hypoglossal nerve (XII)

Cribiform plate

Skeleton   3
TA Latin: *Lamina cribrosa ossis ethmoidalis*
TA English: *Cribiform plate of ethmoid*
A small bone plate that is greatly perforated, so that the fibers of the olfactory nerve (I) can penetrate it and reach the olfactory epithelium in the nasal cavity.

Crista galli

Medulla spinalis   2
TA Latin: *Crista galli*
TA English: *Crista galli*
Crista galli. Bony elevation in the middle of the ethmoid bone, from which the falx cerebri stretches to the internal occipital protuberance.

Crural cistern

Meninges & Cisterns   2
Joins the cisterna ambiens laterally and is also often viewed as being part of the latter.
It forms the space between the crus cerebri and the parahippocampal gyrus and contains the choroid artery.

Crus of fornix

Diencephalon   1
TA Latin: *Crus fornicis*
TA English: *Crus of fornix*
Part of the fornix. In this section, the fibers of the alveus unite before passing – to a small extent – via the commissure of the fornix to the contralateral side. The larger part bundles with the contralateral fibers forming the body of fornix and passes to the hypothalamus.

Crystalline lens

Eye   2
TA Latin: *Lens*
TA English: *Lens*
Lens of the eye.

CSA (common somato-afferent series)

General CNS   1
The nuclei of the cranial nerves can be categorized as per Gaskell and Herrick into seven longitudinally oriented columns:
– GSA: general somato-afferent series
– GSE: general somato-efferent series
– GVA: general viscero-afferent series
– GVE: general viscero-efferent series
– SVA: special viscero-afferent series
– SVE: special viscero-efferent series
– SSA: special somato-afferent series

Culmen

Cerebellum   3
TA Latin: *Culmen*
TA English: *Culmen*
Part of the vermis cerebelli lying above the primary fissure. Belongs to the anterior lobe.
Like the entire vermis cerebelli, the culmen receives its afferents primarily from the spinal cord. Hence it is part of the so-called spinocerebellum = palaeocerebellum.

Cuneate fasciculus

Medulla spinalis   3
TA Latin: *Fasciculus cuneatus*
TA English: *Cuneate fasciculus*
The cuneate fasciculus and gracile fasciculus together form the posterior column and are the main axes of epicritic sensibility.
– gracile fasciculus: it collects the epicritic fibers from the sacral, lumbar as well as lower thoracic cord and terminates in the gracile nucleus.
– cuneate fasciculus: contains the fibers from the upper thoracic cord as well as from the cervical cord and terminates in the cuneate nucleus.

Cuneiform nucleus

Mesencephalon   1

TA Latin: *Nucl. cuneiformis*
TA English: *Cuneiform nucleus*
It is important to bear in mind that the mesencephalic cuneiform nucleus and sub-cuneiform nucleus as well as a narrow medial strip of the central myelencephalon nucleus are part of the medial zone of the reticular formation.

Cuneocerebellar tract
Cerebellum 2
The posterior cuneocerebellar tract conducts primary afferents from the cuneate nucleus to the cerebellum.
The mossy fibers of this tract have their origin in the lateral cuneate nucleus and conduct proprio– and exteroceptive impulses from the upper limbs without synapsing in the cerebellar hemispheres.

Cuneus
Telencephalon 2
TA Latin: *Cuneus*
TA English: *Cuneus*
Part of the occipital lobe visible on the medial aspect of the hemisphere. Involved in the processing of visual information.

Cytoarchitectonic fields of Brodmann
Telencephalon 3
Korbinian Brodmann (1868–1918) studied the structure of the cerebral cortex and, based on cytoarchitectonic considerations, divided it into areas to which he assigned numbers. Accordingly, e.g. area 17 corresponds to the striate cortex or area 4 to the motor cortex.

D

Darkschewitsch's nucleus

Mesencephalon 1

TA Latin: *Nucl. commissurae post. (Darkschewitsch)*

TA English: *Nucleus of posterior commissure (Darkschewitsch)*

Darkschewitsch´s nucleus and the interstitial nucleus (Cajal) are two groups of small cells within the reticular formation. Afferents arise from the ipsilateral corpus striatum and globus pallidus, vestibular nuclei as well as the contralateral cerebellum. Efferents pass across the interstitiospinal tract into the spinal cord. This nucleus is part of the motor system (oculomotor) and shares responsibility for regulation of muscle tone.

Declive

Cerebellum 3

TA Latin: *Declive*

TA English: *Declive*

Part of the vermis cerebelli lying below the primary fissure. Belongs to the posterior lobe.

Like the entire vermis cerebelli, the declive receives its afferents primarily from the spinal cord. Hence it is part of the so-called spinocerebellum = palaeocerebellum.

Decussation of the lateral lemnisci

Mesencephalon 2

In the decussation of the lateral lemnisci, the efferents of the lateral lemniscal nucleus decussate to the contralateral side and terminate there:

– in the lateral lemniscal nucleus of the contralateral side

– in the contralateral inferior colliculus.

This decussation can also be called decussation of the internal arcuate fibers.

Decussation of the superior cerebellar peduncles

Cerebellum 2

TA Latin: *Decussatio pedunculorum cerebellarium sup.*

TA English: *Decussation of superior cerebellar peduncles*

The fibers of the superior cerebellar peduncle passing in the direction of the thalamus and red nucleus decussate to the contralateral side in the tegmentum of mesencephalon. This decussation site of these large fiber bundles is called the decussation of the superior cerebellar peduncles.

Decussation of the trigeminothalamic tracts

Mesencephalon 1

These fibers are efferent fibers of the principal nucleus of the trigeminal nerve. After exiting from the nucleus, these decussate to the contralateral side (ventral tegmental fasciculus), bundle there to form the trigeminal lemniscus and then ascend to the ventral posteromedial thalamic nucleus. They conduct somatosensory information from the entire facial surface.

Decussation of the trochlear nerves

Diencephalon 1

In the rostral part of the superior medullary velum, the efferents of the nucleus of the trochlear nerve decussate to the contralateral side. This is called the decussation of the trochlear nerves. The fibers emerge from the mesencephalon directly beneath the inferior colliculus and form the trochlear nerve (IV), the only cranial nerve to emerge from the dorsal side of the brainstem.

Decussation of the uncinate cerebellar fasciculi

Cerebellum 1

The uncinate cerebellar fasciculus decussates to the contralateral side immediately after exiting from the fastigial nucleus, then passing on to the vestibular nuclei. This decussation site is called decussation of the uncinate cerebellar fasciculi.

Deep cervical artery

Vessels 2

TA Latin: *A. cervicalis profunda*

TA English: *Deep cervical artery*

Arises from the costocervical trunk and supplies vertebral canal, medullary cistern, spinal cord as well as the deep neck and cervical muscles.
The trunk can also be the origin of the radicular arteries.

Deep cervical vein
Vessels   2
TA Latin: *V. cervicalis profunda*
TA English: *Deep cervical vein*
It carries blood from the deep neck and cervical musculature as well as from the occipital vein to the brachiocephalic vein.

Deep facial vein
Vessels   1
TA Latin: *V. profunda faciei*
TA English: *Deep facial vein*
Deep facial veins forming an anastomosis between the facial vein and pterygoid sinus.

Deep gray layer of the superior colliculus
Mesencephalon   1
TA Latin: *Stratum griseum profundum colliculi sup.*
TA English: *Deep grey layer of superior colliculus*
→ Superior colliculus, deep gray layer

Deep interpeduncular branches
Vessels
TA Latin: *Aa. centrales posteromed.*
TA English: *Posteromedial central arteries*
→ Posteromedial central arteries

Deep middle cerebral vein
Vessels   2
TA Latin: *V. media profunda cerebri*
TA English: *Deep middle cerebral vein*
The deep middle cerebral vein collects venous blood from the insular region, unites with the anterior cerebral vein, thus creating the basal vein.

Deep white matter of the superior colliculus
Mesencephalon
TA Latin: *Colliculus sup., Stratum medullare profundum*
TA English: *Superior colliculus, deep white layer*
→ Superior colliculus, deep white layer

Dementia
In the presence of presenile or senile demence, especially in the case of Alzheimer's disease, there is marked degeneration of the cells of the basal nucleus Meynert (Ch.4), with concomitant impairment of memory, disorientation, motor unrest and impaired speech.

Dentate fascia
Telencephalon   1
→ Dentate gyrus

Dentate gyrus
Telencephalon   3
TA Latin: *Gyrus dentatus*
TA English: *Dentate gyrus*
The dentate gyrus (dentate fascia) is an important part of the hippocampus, retrocommissural part. Rostrally, it joins Giancomini's band and caudally, the fasciolar gyrus. Afferents come from the hypothalamus and via the entorhinal area from the cerebral cortex. Efferents of the granular cells pass exclusively to the cell layers CA3 and CA4 of Ammon´s horn, forming the dense mossy-fiber system.

Dentate nucleus
Cerebellum   3
TA Latin: *Nucl. dentatus*
TA English: *Dentate nucleus*
Measuring about 1 cm in length, the dentate nucleus is the largest cerebellar nucleus. With its typical, saw-toothed shape it lies in the medulla of the cerebellar hemisphere. Afferents: (1) Purkinje fibers of the cerebellar hemisphere, lateral part, (2) collaterals of the pontocerebellar projection.
Efferents: via superior cerebellar peduncle to the ventral lateral thalamic nucleus. A few fibers terminate in the red nucleus.

Dermatome
A spinal nerve originates in the spinal cord, unlike the cranial nerve which arises from the cerebrum. A distinction is made between 31 pairs: 8 cervical (C1–C8), 12 thoracic (Th1–Th12), 5 lumbar (L1–L5), 5 sacral (S1–S5), and 1 coccygeal. Each segment pair provides sensory innervation for a clearly delineated skin area (=dermatome). Close to the spinal cord, the spinal nerve divides into a sensory dorsal root and a motor ventral

root. External to the intervertebral foramen it divides again into a ventral branch and a dorsal branch.

Descending conjunctive brachium
Cerebellum
→ Superior cerebellar peduncle, descending branch

Descending venule
Vessels   1
Small veins descending to the posteromedian medullary vein.

Déviation conuguée
→ Superior frontal gyrus

Diabetes insipidus
→ Supraoptic nucleus

Diagonal band (Broca)
Diencephalon   2
TA Latin: *Stria diagonalis (Broca)*
TA English: *Broca´s diagonal band*
Part of the septal verum, frontier to the anterior perforated substance.
Receives afferents from the lateral septal nucleus, mammillary body, medial nucleus and dorsal tegmental nucleus.
Efferents to the olfactory bulb, hippocampus and entorhinal cortex as well as to the preoptic area, lateral hypothalamic area, mammillary body and to the raphe nuclei of the mesencephalon.

Diagonal gyrus
Telencephalon   2
Continuation of the paraterminal gyrus. It runs along the border between anterior perforated substance and thalamus. At a deep level, it contains important thalamic nuclei such as the medial thalamic nucleus and red nucleus.

Diaphragma sellae
Meninges & Cisterns   1
TA Latin: *Diaphragma sellae*
TA English: *Diaphragma sellae*
Part of the dura mater. Rests on the sella turcica and spreads over the hypophysis.

Diencephalic branches (inferior)
Vessels

→ Inferior thalamic branch

Diencephalic branches (posterior and superior)
Vessels

→ Posterior cerebral artery, postcommunical part, thalamic branches

Diencephalon
Diencephalon   3
TA Latin: *Diencephalon*
TA English: *Diencephalon*
The highest part of the brainstem, contains the tissue surrounding the third ventricle. To it belong the hypothalamus and the dorsal thalamic regions contiguous with the hypothalamus.
Autonomic centers (hormone center) for regulating meolism, heat and water balance, blood pressure and sweat secretion are encountered here as well as in the thalamic nuclear regions of the extrapyramidal system (e.g. globus pallidus).

Digitationes hippocampi
Telencephalon   1
TA Latin: *Digitationes hippocampi*
TA English: *Hippocampal digitations*
Fingerlike evaginations of Ammon's horn.

Diploic veins
Vessels   2
TA Latin: *Vv. diploicae*
TA English: *Diploic veins*
The diploic veins are veins that course in the spongy matter of the calvaria.

Dopaminergic cell groups A8–A10   1
TA Latin: *Cellulae dopaminergicae (A8–A10)*
TA English: *Dopaminergic cells (A8–A10)*
The dopaminergic cell groups A8 to A10 are located in the mesencephalon.
A8 and A9 together form the nigrostratial dopaminergic system forming major efferents across the hypothalamus and internal capsule to the caudate nucleus and putamen and facilitating motor programs. The dopaminergic cell group A10 forms the mesolimbic dopaminergic system sending efferents via the medial forebrain bundle to the limbic system.
Dopamine deficit produces hypokinetic symp-

toms in the motor system, whereas a surfeit results in hyperkinesis.

Dorsal accessory nucleus of the olive 2
TA Latin: *Nucl. olivaris accessorius post.*
TA English: *Posterior accessory olivary nucleus*
The two accessory olives (dorsal accessory nucleus of the inferior olive and medial accessory nucleus of the inferior olive) receive afferents from the central gray matter of mesencephalon and the spinal cord (via spino-olivary tract) and project to the cerebellum, to the following parts: interpositus nucleus (emboliform + globose nuclei), fastigial nucleus, vermis cerebelli and cerebellar hemisphere, intermediate part.
The accessory olives are also involved in movement coordination.

Dorsal acoustic stria
Myelencephalon 3
Both cochlear nuclei project to the contralateral inferior colliculus.
The ventral cochlear nuclei do so via the trapezoid body, while the dorsal cochlear nuclei employ the dorsal acoustic stria to this effect. The latter cross to the contralateral side, passing from there as the lateral lemniscus to the inferior colliculus.

Dorsal ascending serotoninergic pathway
Pathways 1
The dorsal tract is composed of fibers of the dorsal raphe nucleus but also of the superior central nucleus and of the raphe magnus nucleus. Target regions are the gray matter of the mesencephalon and the posterior hypothalamic area. The fibers travel initially with the dorsal longitudinal fasciculus (Schütz), then with the medial forebrain bundle.

Dorsal branch of corpus callosum
Vessels 1
TA Latin: *R. corporis callosi dors.*
TA English: *Dorsal branch to corpus callosum*
A commonly encountered lateral branch of the middle occipital artery.

Dorsal branch of the 8th thoracic nerve
Nerves 1
TA Latin: *R. post. n. thoracicorum (VIII)*
TA English: *Posterior ramus of 8th thoracic nerve*
Having exited from the trunk of the thoracic nerve, the nerve divides, as is typical of spinal nerves, into a ventral branch and a dorsal branch.

Dorsal branch of the hemiazygos vein
Vessels 2
The dorsal branch of the hemiazygos vein can flow into the intercostal vein.

Dorsal cochlear nucleus
Myelencephalon 3
TA Latin: *Nucl. cochlearis post.*
TA English: *Posterior cochlear nucleus*
→ Cochlear nuclei

Dorsal descending serotoninergic pathway
Pathways 1
The descending pathways of serotoninergic cells pass to the cerebellum (from raphe obscurus nucleus (B2) and raphe pontine nucleus (B5)), locus coeruleus (from raphe magnus nucleus (B3)) to several nuclei in the pons and medulla, right down to the spinal cord (from groups raphe pallidus nucleus (B1), B2 and B3), where they terminate predominantly in the anterior column and in the intermediolateral nucleus.

Dorsal horn of spinal cord
Medulla spinalis
TA Latin: *Cornu posterius*
TA English: *Posterior horn of spinal cord*
→ Posterior horn

Dorsal lateral thalamic nucleus
Diencephalon 2
TA Latin: *Nucl. lat. dors. thalami*
TA English: *Dorsal lateral thalamic nucleus*
This thalamic nucleus of the lateral nuclear group is, like the anterior thalamic nucleus, reciprocally connected with the limbic cortex of the cingulate gyrus, retrosplenial area as well as the pre- and parasubiculum. Concomitantly, it receives afferents from the pretectal area and projects to the hippocampus, parietal lobe and retrosplenial cortex and is involved in somatosensory-motor integration processes.

Dorsal longitudinal fasciculus, (Schütz)

Diencephalon   3

TA Latin: *Fasciculus longitudinalis post.*

TA English: *Posterior longitudinal fasciculus*

The dorsal longitudinal fasciculus is a central axis of the autonomic nervous system, coupling the hypothalamus to the nuclei of the brainstem, primarily the parasympathetic nuclear regions, cranial nerve nuclei of the vagus nerve (X), trigeminal nerve (V), hypoglossal nerve (XII) and facial nerve (VII). Ascending fibers come from the solitary nucleus and the reticular formation, conveying predominantly gustatory information to the hypothalamus.

Dorsal nasal artery

Vessels   2

TA Latin: *A. dors. nasi*

TA English: *Dorsal nasal artery*

Artery emerging from the facial artery, which courses with the supratrochlear artery in the inner angle of eye and is a component of the extracranio-orbital anastomosis.

Dorsal noradrenergic bundle, caudal limb

Mesencephalon   3

Around half of all noradrenergic neurons are located in the locus coeruleus. With their transmitter noradrenaline, they generate an inhibitory effect on the cells activated by acetylcholine. The caudal, noradrenergic bundle consists of the descending efferents of the locus coeruleus. These pass into the brainstem, pons and spinal cord. (cf. also "dorsal noradrenergic bundle, rostral limb")

Dorsal noradrenergic bundle, rostral limb

Mesencephalon   3

Around half of all noradrenergic neurons are located in the locus coeruleus. With their transmitter noradrenaline, they generate an inhibitory effect on the cells activated by acetylcholine. The dorsal noradrenergic bundle comprises the ascending efferents of the locus coeruleus. They pass into the entire diencephalon, limbic system and cerebellum. (cf. also "dorsal noradrenergic bundle, rostral limb")

Dorsal nucleus of the vagus nerve

Myelencephalon   3

TA Latin: *Nucl. post. n. vagi*

TA English: *Posterior nucleus of vagus nerve*

In this approximately 2 cm long nucleus, the preganglionic parasympathetic (GVE) fibers originate. The nucleus runs in the medulla somewhat parallel to the nucleus of the hypoglossal nerve, in the lower angle of the fourth ventricle (vagal trigone).

Afferents are received by the nucleus from the solitary nucleus, dorsal tegmental nucleus and other structures.

The special visceromotor fibers in the vagus nerve originate in the nucleus ambiguus, as in the case of those from glossopharyngeal nerve (IX).

Dorsal premammillary nucleus

Diencephalon   1

TA Latin: *Nucl. premammillaris dors.*

TA English: *Dorsal premammillary nucleus*

The premammillary nuclei form the medial hypothalamica area. The dorsal nucleus receives afferent fibers from the ventromedial hypothalamic nucleus and the lateral hypothalamic area. The ventral premammillary nucleus is connected with the amygdala.

Both nuclei are involved in offensive and defensive behaviours.

Dorsal raphe nucleus (B7)

TA Latin: *Nucl. raphes post. (B7)*

TA English: *Posterior raphe nucleus (B7)*

→ Raphe nuclei

Dorsal raphespinal projection

Pathways   1

Efferent fibers of the raphe nucleus, particularly from the raphe magnus nucleus (B3), raphe pallidus nucleus (B1) and raphe obscurus nucleus (B2), go as far as the gray mater of the spinal cord.

Dorsal root of the spinal nerve

Medulla spinalis   3

TA Latin: *N. spinalis, radix post.*

TA English: *Posterior root of the spinal nerve*

Via the dorsal root, peripheral sensory nerve fibers enter the spinal cord. A distinction is made between various types of fibers (A, C fibers etc.)

Dorsal sacral foramina

Meninges & Cisterns   1

TA Latin: *Foramina sacralia posteriora*
TA English: *Posterior sacral foramina*
A window in the sacral bone, opening dorsally.

Dorsal striatum

Telencephalon 1
TA Latin: *Striatum dorsale*
TA English: *Dorsal striatum*
Dorsal part of the corpus striatum.

Dorsal tegmental decussation (Meynert)

Mesencephalon 1
TA Latin: *Decussatio tegmentalis post.*
(Meynert)
TA English: *Posterior tegmental decussation*
(Meynert)
In the dorsal tegmental decussation (named for its discoverer Meynert) the tectospinal tract decussates to the contralateral side. Here fibers run from the superior colliculus to the upper cervical segments, where they terminate in Rexed's laminae VI to VIII.

Dorsal tegmental fasciculus (Shute and Lewis)

Mesencephalon 1
Important efferent of the cholinergic cell groups. It connects these with the superior colliculus and with various nuclei of the thalamus such as the intralaminar nuclei.

Dorsal tegmental nucleus (Gudden)

Mesencephalon 2
TA Latin: *Nucl. tegmentalis post.*
TA English: *Posterior tegmental nucleus*
Cell cluster in the quadrigeminal plate. Belongs to the limbic midbrain region and features afferents from the mammillary body, raphe nuclei, habenular nucleus and interpeduncular nucleus. Efferents go via the dorsal longitudinal fasciculus, inter alia, to the mammillary body and posterior hypothalamic nucleus.

Dorsal terminal nucleus

Diencephalon 1
TA Latin: *Nucl. post. accessorii tracti optici*
TA English: *Posterior nucleus of accessory nuclei of optic tract*
Nuclear region of the superior colliculus which merges with the optic tract nucleus.
Is a component of the accessory optic system

and is involved in coupling of visual information and head movement.

Dorsal thalamus

Diencephalon 3
TA Latin: *Thalamus*
TA English: *Thalamus*
The diencephalon comprises:
– thalamus (dorsal thalamus)
– epithalamus,
– metathalamus,
– subthalamus,
– hypothalamus.

Dorsal trigeminothalamic tract (Wallenberg)

Myelencephalon 3
TA Latin: *Tractus trigeminothalamicus post.*
TA English: *Posterior trigeminothalamic tract*
From the principal nucleus of the trigeminal nerve there are two large efferent fiber bundles: the dorsal trigeminothalamic tract goes to the ipsilateral thalamic nucleus, while the dorsal tegmental fasciculus goes to the contralateral ventral posteromedial thalamic nucleus. This nuclear region projects in turn to the postcentral gyrus of the cerebral cortex, i.e. into the somatosensory cortex. Information is conducted, inter alia, from the skin areas of forehead and face.

Dorsal vagus complex 1

The dorsal vagus complex consists of the solitary nucleus and the dorsal nucleus of the vagus nerve, i.e. both visceral nuclei of the vagus nerve (X).

Dorsal vein of the corpus callosum

Vessels 2
TA Latin: *V. posterior corporis callosi*
TA English: *Posterior vein of corpus callosum*
The dorsal vein of the corpus callosum courses around the splenium of the corpus callosum and ascends to the roof of the corpus callosum, up to about the level of the central sulcus.
Carries venous blood into the inferior cerebral vein.

Dorsolateral fasciculus of spinal cord (Lissauer)

Medulla spinalis 2
TA Latin: *Tractus posterolat. (Lissauer)*

TA English: *Posterolateral tract (Lissauer)*
Like the fasciculus proprius, Lissauer´s tract contains primarily fibers for the intrinsic and reflex apparatus of the spinal cord.

Dorsolateral tegmental fasciculus
Pathways   2
TA Latin: *Tractus trigeminothalamicus post.*
TA English: *Posterior trigeminothalamic tract*
→ Dorsal trigeminothalamic tract (Wallenberg)

Dorsomedial hypothalamic nucleus
Diencephalon   2
TA Latin: *Nucl. dorsomed. hypothalami*
TA English: *Dorsomedial hypothalamic nucleus*
A diffusely organized hypothalamic nucleus implicated in eating behavior. Afferents from many subcortical areas. Efferents to the paraventricular nucleus, parvocellular part (influences neuorendocrine system).
The motor nucleus of the vagus nerve (parasym. effect on endocrine pancreas ⇒ insulin production), circumventricular organs (control of humoral factors from the blood).

Dorsomedial nucleus of hypothalamus
Diencephalon
TA Latin: *Nucl. dorsomed. hypothalami*
TA English: *Dorsomedial nucleus of hypothalamus*

→ Dorsomedial hypothalamic nucleus

Dorsum sellae
Skeleton   1
TA Latin: *Dorsum sellae*
TA English: *Dorsum sellae*
Rear portion of the sella turcica, with the hypophysis situated at a deep level.

Dura mater
Meninges & Cisterns   3
TA Latin: *Dura mater*
TA English: *Dura mater*
The dura mater belongs to the three cranial meninges. As hard meninges, it is compared with the soft meninges consisting of the two other meninges, arachnoid and pia mater.
In the dura mater of the brain it can hardly be separated form the calvaria periosteum, in the spinal dura mater it is separated from the periosteum by means of the epidural cavity.

Dura mater of brain
Meninges & Cisterns   3
TA Latin: *Dura mater cranialis*
TA English: *Cranial dura mater*
Pachymeninx. One of the three meninges. In the dura mater of brain it is closely connected with the periosteum of the calvaria, lined with the arachnoid on the inside. Consists of two layers which spread out along the longitudinal fissure of cerebrum the falx cerebri (falx = sickle), and between cerebrum and cerebellum the tentorium cerebelli. Also forms blood sinuses such as the superior sagittal sinus and inferior sagittal sinus.

Dural sac of the spinal ganglion and spinal roots
Meninges & Cisterns   3
Evaginations of the spinal dura mater surrounding the roots and ganglia of the spinal nerves.

Dystonic syndrome
Damage to the corpus striatum results in the typically manifest symptoms of chorea, due to disinhibition of the globus pallidus and substantia nigra. Chorea is characterized at an advanced stage by hyperkinesia especially of the distal extremities' musculature and of the face. Dystonic syndrome (e.g. retrocollis, spastic torticollis) or athetosis are also encountered.

E

Edinger–Westphal nucleus

TA Latin: *Nuclei viscerales (Edinger-Westphal)*
TA English: *Visceral nuclei (Edinger-Westphal)*
→ Accessory nucleus of oculomotor nerve

Emboliform nucleus

Cerebellum   3
TA Latin: *Nucl. interpositus ant.*
TA English: *Anterior interpositus nucleus*
The emboliform nucleus and globose nucleus are collectively known as the interpositus nucleus, since both receive their afferents from Purkinje cells of the cerebellar hemisphere, intermediate part and from collateral spino- and rubrocerebellar projections. Their efferents pass via the fiber bundle of the superior cerebellar peduncle primarily to the small-celled red nucleus, but with fewer going to the ventral lateral thalamic nucleus.

Emissary veins

Vessels   2
TA Latin: *Vv. emissariae*
TA English: *Emissary veins*
The emissary veins are connections between the sinus of the dura mater, diploic veins and superficial cranial veins.

Encephalon

General CNS   3
TA Latin: *Encephalon*
TA English: *Brain*
Encephalon comprises the part of the CNS located in the calvaria.

Encephalopathy, alcoholic

Damage to the mammillary body, e.g. in the case of alcoholic encephalopathy, results in affective impairments and marked loss of perceptivity.

Endosteal layer of the dura mater

Meninges & Cisterns   1
The layer of dura mater directly connected with the periosteum is called the endosteal layer of dura mater.

Endothelium

Meninges & Cisterns   1
TA Latin: *Endothelium*
TA English: *Endothelium*
Single-layer lining of vessels or serous cavities.

Entorhinal area (entorhinal cortex)

Telencephalon
The entorhinal and perirhinal cortices are situated in the parahippocampal gyrus and mark the transition from allocortex of the hippocampus to the cerebral cortex of the temporal lobe. The area stretches from the amygdaloid body to the prepiriform cortex and already features a 6-layered structure. Afferents: rhinencephalon, Ammon's horn, septum verum, cortex, thalamus.
Efferents: hippocampus, thalamus, tegmentum of mesencephalon.

Entorhinal cortex (Area 28)

Telencephalon
→ Entorhinal and perirhinal cortices

Entorhinal and perirhinal cortices

Telencephalon   2
The entorhinal and perirhinal cortices are situated in the parahippocampal gyrus and mark the transition from allocortex of the hippocampus to the cerebral cortex of the temporal lobe. The area stretches from the amygdaloid body to the prepiriform cortex and already features a 6-layered structure. Afferents: rhinencephalon, Ammon's horn, septum verum, cortex, thalamus.
Efferents: hippocampus, thalamus, tegmentum of mesencephalon.

Epicritic sensibility

As regards the somatosensory control, there are two groups of sensibility:
– the protopathic sensibility includes crude touch and pressure perceptions, pain and temperature.

– the epicritic sensibility comprises exteroceptive stimuli (exact, tactile stimuli of the
mechanoreceptors of the skin) and
proprioceptive stimuli (position of the body
in space, mediated by joint and muscle receptors).

Epidural cavity

Meninges & Cisterns   3
TA Latin: *Spatium epidurale*
TA English: *Epidural space*
By virtue of the fact that the spinal dura mater is
not fused with the periosteum of the vertebral
bodies, a space called the epidural cavity is
formed. This spreads across the entire length of
the vertebral column. The cervical, thoracic and
lumbar epidural cavity is globally known as the
peridural cavity. This plays an important role in
peridural anesthesia.

Epithalamus

Diencephalon   3
TA Latin: *Epithalamus*
TA English: *Epithalamus*
Part of diencephalon. Consists of pineal body
and habenular nuclei.

External acoustic pore (porion), upper edge

Skeleton   1
TA Latin: *Porus acusticus externus*
TA English: *External acoustic pore*
Upper margin of the external acoustic meatus.

External arcuate fibers

Pathways
TA Latin: *Fibrae arcuatae externae*
TA English: *External arcuate fibres*
The arcuate nuclei are situated on the ventral
surface of the pyramid. From here fibers course
as external arcuate fibers via the superior cerebellar peduncle to the cerebellum. Other fibers
transverse the brainstem and, as medullary
striae on the floor of the fourth ventricle, they
reach the inferior cerebellar peduncles, via
which they pass into the cerebellum.

External capsule

Telencephalon   3
TA Latin: *Capsula externa*
TA English: *External capsule*

The external capsule is a fiber layer running between claustrum and putamen.
It contains projection fibers from the frontoparietal operculum and other parietal cortical
areas.

External carotid artery

Vessels   3
TA Latin: *A. carotis externa*
TA English: *External carotid artery*
Emerges with the internal carotid artery from
the division of the common carotid artery. It
gives off important extracranial branches, such
as the facial artery, occipital artery and ascending pharyngeal artery, before dividing into the
superficial temporal artery and the maxillary artery.

External iliac vein

Vessels   3
TA Latin: *V. iliaca ext.*
TA English: *External iliac vein*
Drains venous blood from the leg, abdominal
wall and lumbar region into the common iliac
vein.

External jugular vein

Vessels   3
TA Latin: *V. jugularis externa*
TA English: *External jugular vein*
Jugular veins that run close to the surface and
drain venous blood from the maxillary vein, superficial temporal veins, occipital region, neck
and shoulders into either the internal jugular
vein or directly into the subclavian artery.

External maxillary artery

Vessels
TA Latin: *A. facialis*
TA English: *Facial artery*
→ Facial artery

External medullary lamina of the thalamus

Diencephalon   1
TA Latin: *Lamina medullaris externa*
TA English: *External medullary lamina of
thalamus*
Border layer between thalamus and internal
capsule.

External occipital protuberance
Skeleton 1
TA Latin: *Protuberantia occipitalis externa*
TA English: *External occipital protuberance*
Bony protuberance in the center of the occipital bone that can be easily felt from outside. Lies just above the insertion of the neck muscles, in a direct prolongation of the spiny processes of the vertebral column.

Exteroceptive stimuli
As regards the somatosensory control, there are two groups of sensibility:
- the protopathic sensibility includes crude touch and pressure perceptions, pain and temperature.
- the epicritic sensibility comprises exteroceptive stimuli (exact, tactile stimuli of the mechanoreceptors of the skin) and proprioceptive stimuli (position of the body in space, mediated by joint and muscle receptors).

Extra-ocular muscles
Eye 1
TA Latin: *Musculi externi bulbi oculi*
TA English: *Extra-ocular muscles*
The extra-ocular muscles are compared with the intra-ocular muscles.
The former provide for movement of the eyeball, while the latter are involved in adaptation (via the iris) and accommodation (via lens).

Extracranio-orbital cerebromeningeal anastomosis
Vessels 2
Anastomoses between superficial arteries, which are close to the orbita, and vessels that are near the meninges.
Such anastomoses may be implicated in pathogenic dissemination.

Extrageniculate visual tract
Nerves 2
The term extrageniculate is used to designate the sum of all tracts which caudal to the optic chiasm do not project to the LGB.
These include the retinohypothalamic projections to the suprachiastmatic nucleus as well as the optic tract, lateral root with its accessory optic tract branching to the terminal nuclei.

Extreme capsule
Telencephalon 2
TA Latin: *Capsula extrema*
TA English: *Extreme capsule*
Laterally, the claustrum is differentiated from the insula (of Reil) by virtue of a lamina of white mater, the extreme capsule.

pression are manifest even including complete paralysis of the ipsilateral facial musculature.

F

Facial artery

Vessels   3

TA Latin: *A. facialis*

TA English: *Facial artery*

Like the maxillary artery, the facial artery arises from the external carotid artery and supplies with its branches the upper pharynx, palate, palatine tonsils, muscles and mucosa of the lips, nasal septum, alae of the nose, dorsal nose and various facial muscles.

Facial colliculus

Pons   2

TA Latin: *Colliculus facialis*

TA English: *Facial colliculus*

Tuberculum on the floor of the fourth ventricle caused by the fibers of the facial nerve coursing over the abducens nucleus.

Facial nerve (VII)

Nerves   3

TA Latin: *N. facialis (N.VII)*

TA English: *Facial nerve (VII)*

The facial nerve conducts three qualities:

1) motor efferents for innervating the mimetic muscles:

Nucleus: nucleus of the facial nerve.

2) visceromotor control: parasympathetic innervation of salivary and lacrimal glands.

Nucleus: salivatory nuclei.

3) Somatosensory control: sensory innervation of the tongue (anterior 2/3) and of the external ear.

Nucleus: solitary nucleus.

Qualities (2) and (3) are mediated by the intermediate maxillary nerve (V2) which is a permanent part of the facial nerve.

Skull: internal acoustic meatus.

Peripheral damage to the facial nerve leads to facial paresis. Depending on the localization of the lesion, symptoms with different degrees of ex-

Facial paralysis

→ Facial nerve (VII)

Facial vein

Vessels   3

TA Latin: *V. facialis*

TA English: *Facial vein*

Large facial vein, arising from the angular vein. It collects blood from the soft tissues and muscles of the face, temporomandibular joint, parotid gland and dura mater.

Carries the blood to the internal jugular vein, from which it flows via the brachiocephalic vein into the superior vena cave.

Falx cerebelli

Cerebellum   2

TA Latin: *Falx cerebelli*

TA English: *Falx cerebelli*

The tuber vermi and pyramid vermis are located quite deep between the two cerebellar hemispheres. Into this cleft, resting on the occipital bone, protrudes an evagination of dura mater, which, shaped like a sickle, follows the course of this groove. This sickle-shaped evagination of the dura mater is called the falx cerebelli. Its bigger counterpart is called the falx cerebri. This divides the two cerebral hemispheres.

Falx cerebri

Meninges & Cisterns   3

TA Latin: *Falx cerebri*

TA English: *Falx cerebri*

The dura mater forms a tough, sickle-shaped (falx = sickle) layer of tissue that stretches over the entire length and depth of the longitudinal fissure of cerebrum.

At the cranium it forms a cavity, the superior sagittal sinus, and likewise at its free end: inferior sagittal sinus.

Fasciculus

General CNS   3

TA Latin: *Fasciculus*

TA English: *Fasciculus*

Longitudinal bundle.

Fasciculus proprius

Medulla spinalis 2
TA Latin: *Fasciculi proprii*
TA English: *Fasciculus proprius*
This is a border of white matter around the H-shaped gray matter of the spinal cord. In this border run short fibers, forming the intrinsic and reflex apparatus of the spinal cord and endowed with the ability to stretch over several segments.

Fasciculus retroflexus

Pathways
TA Latin: *Tractus habenulointerpeduncularis (Meynert)*
TA English: *Habenulo-interpeduncular tract (Meynert)*
→ Habenulo-interpeduncular tract (Meynert)

Fasciola cinera

Telencephalon 1
→ Fasciolar gyrus

Fasciolar gyrus

Telencephalon 1
TA Latin: *Gyrus fasciolaris*
TA English: *Fasciolar gyrus*
At the fasciola cinera, the dentate gyrus joins the fasciolar gyrus via which it has contact with the indusium griseum.

Fastigial nucleus

Cerebellum 3
TA Latin: *Nucl. fastigii*
TA English: *Fastigial nucleus*
The fastigial nucleus receives its afferents (1) from the Purkinje fibers of the vermis cerebelli, (2) directly from the vestibular apparatus and (3) via collaterals from the vestibular nucleus. Its efferents course as the uncinate fasciculus via the inferior cerebellar peduncle to the vestibular nuclei and the reticular formation on the contralateral side.

Fastigium

Cerebellum 2
TA Latin: *Fastigium*
TA English: *Fastigium*
The cerebellum rests on the roof the fourth ventricle. Endowed with a sharp gable, this roof penetrates deeply into the trunk of the arbor vitae.

This gable-shaped extension is called the fastigium. It is also an eponymous designation for the fastigial nuclei situated in the cerebellum.

Fastigobulbar tract

Cerebellum 1
Fiber bundles passing from the cerebellar fastigial nucleus along the ventricle wall to the vestibular nuclei.

Fibers of the facial nerve

Pons 1
The fibers emerge directly from the nucleus of the facial nerve (VII) and continuing their course, they go around the nucleus of the abducens nerve in the genu of the facial nerve, before exiting from the brainstem.

Field of vision, binocular part

Eye 2
The field of vision is the spatial area visible to the fixated eye. A distinction is made between monocular and binocular field of vision. The overlapping part of the right and left monocular field is called the central part.

Fila olfactoria

Pathways 3
TA Latin: *Fila olfactoria*
TA English: *Olfactory nerves*
→ Olfactory nerve (I)

Fila radicularia

Medulla spinalis 2
TA Latin: *Fila radicularia*
TA English: *Rootlets*
Fibers emerging from the spinal cord, forming the dorsal roots and ventral roots and uniting to form the spinal nerve.

Filum terminale

Meninges & Cisterns 3
TA Latin: *Filum terminale*
TA English: *Filum terminale*
Thread-shaped terminal segment of the spinal cord. It continues from conus medullaris and runs about 2/3 of the way within the dura mater, with 1/3 on the outside.
Is composed of a mixture of regressed nervous tissue, leptomeninx and connective tissue.

Is formed by a partial regression process during embryonic development.

Fimbria of the hippocampus

Telencephalon   3
TA Latin: *Fimbria hippocampi*
TA English: *Fimbria of hippocampus*
Fibers from the alveus of hippocampus unite to form the fimbria of the hippocampus, and these in turn to form the crus of fornix, which is the initial part of the fornix. Here run efferent fibers from Ammon's horn in the direction of the fornix and via the latter to the thalamic nuclei and mammillary body as well as the septum, nucleus accumbens and hypothalamus.
The tenia of the fornix is also called fimbrial tenia here.

Fimbrial tenia

Meninges & Cisterns   2
Along the fimbria of the hippocampus, the tenia of the fornix is also called the fimbrial tenia.
By means of it, the choroid plexus of the lateral ventricle is attached to the surrounding brain tissue.

Fimbriodentate sulcus

Telencephalon   1
TA Latin: *Sulcus fimbriodentatus*
TA English: *Fimbriodentate sulcus*
Sulcus between the dentate gyrus and the fimbria of the hippocampus.

First lumbar artery

Vessels   1
TA Latin: *A. lumbalis 1*
TA English: *First lumbar artery*
The lumbar arteries arise from the abdominal aorta and supply the vertebral canal, back and abdominal muscles.

First lumbar vein

Vessels   1
TA Latin: *V. lumbalis*
TA English: *First lumbar vein*
The lumbar veins collect venous blood from the back musculature, vertebral canal, spinal cord and abdominal wall, carrying it into the inferior vena cava (3+4) or into the ascending lumbar vein (1+2).

Fissura intrabiventeris

Cerebellum   2
TA Latin: *Fissura intrabiventralis*
TA English: *Intrabiventral fissure*
The fissura intrabiventeris runs transversely through the biventer lobule.

Fissura prebiventeris

Cerebellum   2
TA Latin: *Fissura prebiventralis*
TA English: *Prebiventral fissure*
The fissura prebiventeris separates the gracile lobule and biventer lobule.

Flaccid paralysis

Tumor or hemorrhage in the longitudinal fissure of cerebrum generally triggers symptoms in both body halves. Accordingly, flaccid paralysis of both legs can be induced by a pathological event at the level of area 4.
Damage to the precentral gyrus alone results in flaccid paralysis of the contralateral skeletal musculature. If premotor areas are concurrently affected, spastic paralysis can ensue, as inhibitory influence on the centers of brainstem and thalamus are lacking, hence increased muscle tone of the extrapyramidal system is predominant.

Flechsig pathway

Medulla spinalis
TA Latin: *Tractus spinocerebellaris post. (Flechsig)*
TA English: *Posterior spinocerebellar tract (Flechsig)*
→ Thoracic nucleus

Floccular peduncle

Cerebellum   1
TA Latin: *Pedunculus flocculi*
TA English: *Peduncle of flocculus*
The pedicle-shaped connection between flocculus and tonsil of cerebellum is called the floccular peduncle. It is part of the flocculonodular lobe and receives afferents from the vestibular nuclei.

Flocculonodular lobe

Cerebellum   3
TA Latin: *Lobus flocculonodularis*
TA English: *Flocculonodular lobe*

The vermis segment nodulus and the hemisphere segment flocculus together form the flocculonodular lobe.

Phylogenetically it is very old and is thus called the archicerebellum. Since its afferents come mainly from the vestibular nuclei (vestibulocerebellar tract), the "vestibulocerebellum" is another synonym.

Flocculus

Cerebellum   3
TA Latin: *Flocculus*
TA English: *Flocculus*
→ Flocculonodular lobe

Folium vermis

Cerebellum   2
TA Latin: *Folium vermis*
TA English: *Folium of vermis*
Summit of the vermis cerebelli. It separates the declive and tuber vermis.

Like the entire vermis cerebelli, the declive also receives its afferents primarily from the spinal cord, thus making it part of the so-called spinocerebellum = palaeocerebellum.

Foramen magnum

Meninges & Cisterns   3
TA Latin: *Foramen magnum*
TA English: *Foramen magnum*
The largest hole in the base of the skull. Here the myelencephalon joins the spinal cord. Here is also the transition between brain and spinal cord.

Forel's field   2

TA Latin: *Nuclei campi perizonalis (Forel-Feld)*
TA English: *Nuclei of perizonal fields (Forel's field)*
→ Tegmental area, (Forel's field)

Fornix

Diencephalon   3
TA Latin: *Fornix*
TA English: *Fornix*
Fiber bundles connecting the hippocampus with the hypothalamus and other regions. The hippocampal efferents form the alveus and then unite to form the crus of fornix. Some fibers cross via the commissure of the fornix to the contralateral side, but the majority form bundles with the fibers on the contralateral side, in turn forming the body of fornix, running beneath the corpus callosum, dividing frontally in the columns of fornix and passing to the hypothalamus.

Fourth ventricle

Meninges & Cisterns   3
TA Latin: *Ventriculus quartus*
TA English: *Fourth ventricle*
It lies in the center of the rhombencephalon and stretches caudally into the central canal of the spinal cord. Via the apertures of the fourth ventricle, it releases CSF into the subarachnoid system. The latter is formed by the subarachnoid space and is subdivided into various cisterns. The CSF is reabsorbed by venous blood at the arachnoid granulations of the cranium and at the spinal roots.

Frontal cortex (areas 6+8)

Telencephalon   2
Premotor cortex. It plays a decisive role in the initiation of movements and comprises the frontal eye field (area 8).

Frontal gyri

Telencephalon   3
TA Latin: *Gyri frontales*
TA English: *Frontal gyri*
The frontal gyri form three groups:
– superior frontal gyrus,
– medial frontal gyrus,
– inferior frontal gyrus.
Located in the gyri are also, inter alia, the premotor cortex and the motor speech center (Broca). The prefrontal cortex features additionally areas with pronounced associative functions.

Frontal incision (frontal foramen)

Meninges & Cisterns   2
TA Latin: *Incisura front. (foramen front.)*
TA English: *Frontal notch (frontal foramen)*
Palpable nuclei on the upper margin of the orbita. The supratrochlear artery courses through them.

Frontal lobe

Telencephalon   3

TA Latin: *Lobus front.*
TA English: *Frontal lobe*
The frontal lobe extends from the frontal pole to
the central sulcus.

Frontal oculomotor cortex (area 8)

Telencephalon   1
Frontal eye field. Plays an important role in voluntary control of eye movement.

Frontal operculum

Telencephalon   2
TA Latin: *Operculum frontale*
TA English: *Frontal operculum*
Operculum = lid.
The frontal operculum designates the part of the
frontal lobe covering the insula, which is located
deep in the lateral sulcus.

Frontal pole

Telencephalon   2
TA Latin: *Polus front.*
TA English: *Frontal pole*
Frontal pole of brain on the frontal lobe.

Frontal trunk of the middle cerebral artery

Vessels
→ Middle cerebral artery, frontal trunk

Frontal veins

Vessels   2
TA Latin: *Vv. frontales*
TA English: *Frontal veins*
Frontal veins, frontoparietal veins, parietal
veins and occipital veins form together the superior cerebral veins.
All these veins carry venous blood from their respective catchment areas to the superior sagittal
sinus.

Frontoparietal operculum

Telencephalon   2
Operculum=lid.
The frontoparietal operculum designates the
part of the cerebral cortex at the transition from
frontal lobe to parietal lobe covering the insula,
which is located deep in the lateral sulcus.

Frontopontine tract

Pathways   2
TA Latin: *Tractus frontopontinus*

TA English: *Frontopontine fibres*
Projection of predominantly motor information
from the frontal lobe to the nuclear regions of
the pons (Varolius), where most of them synapse further and go to the cerebellum.

G

Gasserian semilunar ganglion
Nerves
TA Latin: *Ganglion trigeminale (Gasseri)*
TA English: *Trigeminal ganglion (Gasseri)*
→ Trigeminal ganglion

Geniculocalcarine tract
Telencephalon   1
→ Optic radiation

Genu of facial nerve, internal and external
TA Latin: *Genu n. facialis*
TA English: *Genu of facial nerve, internal and external*
→ Genu of the facial nerve

Genu of internal capsule
Telencephalon   3
TA Latin: *Genu capsulae internae*
TA English: *Genu of internal capsule*
→ Internal capsule

Genu of the corpus callosum
Telencephalon   3
TA Latin: *Genu corporis callosi*
TA English: *Genu of corpus callosum*
The frontal lobes of both hemispheres are interconnected via the genu of the corpus callosum. The u-shaped fiber strands are called the frontal forceps.

Genu of the facial nerve
Pathways   1
TA Latin: *Genu n. facialis*
TA English: *Genu of facial nerve*
The facial nerve (VII) embarks on a hairpin-like course around the nucleus of the abducens nerve. This is called the "inner genu of the facial nerve".
The external genu of the facial nerve is situated in the geniculate ganglion in the petrous bone, where the nerve branches off sharply dorsally and laterally.

Giacomini's band
Telencephalon   1
Giacomini´s band is the rostral continuation of the dentate gyrus and divides the uncus into the intralimbic gyrus and uncinate gyrus.

Gigantocellular nucleus
TA Latin: *Nucl. gigantocellularis*
TA English: *Gigantocellular reticular nucleus*
→ Gigantocellular reticular nucleus

Gigantocellular reticular nucleus
Pons   2
TA Latin: *Nucl. reticularis gigantocellularis*
TA English: *Gigantocellular reticular nucleus*
Component of the medial reticular formation. Adjacent to the vestibular complex, it receives afferents from the latter as well as from the fastigial nucleus of the cerebellum and the cerebral cortex. The tectospinal tract brings efferents from the superior colliculus, while the spinoreticular tract conveys signals from the spinal cord.
Efferents go to the locus coeruleus and via the bulbospinal tract to the spinal cord, where they exert an effect on alpha and gamma motoneurons via the interneurons of the intermediate substance.

Glabella
     1
TA Latin: *Glabella*
TA English: *Glabella*
Hairfree section between eyebrows.

Glabella-inion line
Line between glabella and inion. Often used as a reference mark.

Globose nucleus
Cerebellum   3
TA Latin: *Nucl. interpositus post.*
TA English: *Posterior interpositus nucleus*
The emboliform nucleus and globose nucleus are collectively known as the interpositus nucleus, since both receive their afferents from Purkinje cells of the cerebellar hemisphere, intermediate part and from collateral spino- and rubrocerebellar projections. Their efferents

pass via the fiber bundle of the superior cerebellar peduncle primarily to the small-celled red nucleus, but with fewer going to the ventral lateral thalamic nucleus.

Globus pallidus

Diencephalon   3
TA Latin: *Globus pallidus*
TA English: *Globus pallidus*

The globus pallidus belongs to the basal ganglia, with its ontogenetic provenance being the diencephalon (subthalamus). It is subdivided into a globus pallidus, lateral part and a globus pallidus, medial part. Efferents go to the ventral anterolateral thalamic nucleus, afferents to the intralaminar thalamic nuclei, subthalamic nuclei, corpus striatum. Functionally, it acts largely as an antagonist of the striatum, hence facilitating motor information.

By generating a facilitating effect on motor information, dysfunction of the globus pallidus results in hypokinesis and poor timing of movements, thus producing motor clumsiness.

Globus pallidus, lateral part

Diencephalon   2
TA Latin: *Globus pallidus lat.*
TA English: *Globus pallidus lateral segment*

The globus pallidus is subdivided by the medial medullary lamina into:
– globus pallidus, cerebellar hemisphere, lateral part

This constitutes the lateral part of the globus pallidus, where primarily inhibitory fibers of the putamen terminate.
– Globus pallidus, medial part.
   Inner section of the globus pallidus, where fibers of the subthalamic nucleus terminate and from which most efferents exit.

Globus pallidus, medial part

Diencephalon   2
TA Latin: *Globus pallidus med.*
TA English: *Globus pallidus medial segment*
→ Globus pallidus, lateral part

Glossopharyngeal nerve (IX)

Nerves   3
TA Latin: *N. glossopharyngeus (N.IX)*
TA English: *Glossopharyngeal nerve (IX)*

Glossopharyngeal nerve (IX) has no nucleus of its own and contains several fiber types:
– somatomotor control: parts of the pharyngeal muscles.
   Nucleus: nucleus ambiguus
– parasympathetic innervation of the parotid gland.
   Nucleus: salivatory nuclei
– somatosensory control: tongue (posterior 1/3), parts of the pharynx.
   Nucleus: spinal nucleus of the trigeminal nerve.
– viscerosensory control: carotid sinus and glomus caroticum.
   Nucleus: solitary nucleus.
   Skull: jugular foramen.

Gower's band

Medulla spinalis
TA Latin: *Tractus spinocerebellaris ant. (Gower)*
TA English: *Anterior spinocerebellar tract (Gower)*
→ Anterior spinocerebellar tract

Gracile fasciculus

Medulla spinalis   3
TA Latin: *Fasciculus gracilis*
TA English: *Gracile fasciculus*

The cuneate fasciculus and gracile fasciculus together form the posterior column and are the main axes of epicritic sensibility.
– gracile fasciculus: it collects the epicritic fibers from the sacral, lumbar as well as lower thoracic cord and terminates in the gracile nucleus.
– cuneate fasciculus: contains the fibers from the upper thoracic cord as well as from the cervical cord and terminates in the cuneate nucleus.

Gracile lobule

Cerebellum   3
TA Latin: *Lobulus gracilis*
TA English: *Gracile lobule*

The gracile lobule belongs to the posterior lobe and is part of the cerebellar hemispheres. Apart from the areas in proximity to the vermis (intermediate part), the hemispheres belong to the phylogenetically young neocerebellum and receive their afferents via the mossy fibers of the pontocerebellar tract from the pontine nuclei. All hemisphere segments are hence also assigned to the pontocerebellum.

Gracile nucleus
Myelencephalon 3
TA Latin: *Nucl. gracilis*
TA English: *Gracile nucleus*
In the cuneate nucleus and gracile nucleus terminate the epicritic afferents of the posterior column – funiculus dorsalis – (cuneate fasciculus and gracile fasciculus), which is the reason why they are also called posterior column nuclei.
– gracile nucleus: afferents from the trunk and lower extremities.
– cuneate nucleus: afferents from the upper extremities and neck (medial cuneate nucleus) and vestibular organ (lateral cuneate nucleus).
The efferents of both nuclei cross to the contralateral side in the medulla as the internal arcuate fibers and join the trigeminal efferents (epicritic sensibility of the face) to form the medial lemniscus, before passing to the thalamus (ventral posterolateral thalamic nucleus), from where they project into the somatosensory cortex (postcentral gyrus).

Granular cells
Cerebellum 3
Via the mossy fibers they receive afferent impulses from the pontine nuclei, spinal cord and myelencephalon.

Granular foveola (Pacchioni)
Meninges & Cisterns 1
TA Latin: *Foveola granularis (Pacchioni)*
TA English: *Granular foveola (Pacchioni)*
The granular foveola is the name given to a small groove in the calvaria which is created by arachnoid granulations.

Great cerebral vein (Galen)
Vessels 3
TA Latin: *V. magna cerebri (Galeni)*
TA English: *Great cerebral vein (Galen)*
While the superficial cerebral veins drain into the venous blood into the superior sagittal sinus, the deep cerebral veins drain into the great cerebral vein, which enters the straight sinus. The supe-

rior sagittal sinus and straight sinus enter the confluence of the sinuses, situated in the occipital region, with this in turn draining into the transverse sinus and then into the sigmoid sinus. From here, blood enters the internal jugular vein.

Great radicular artery
Vessels 2
In the lumbosacral region the ganglia and nerve roots are supplied by the great radicular artery.

Great spinal artery (Adamkiewicz´s)
Vessels
TA Latin: *A. radicularis ant. (Adamkiewiczi)*
TA English: *Anterior radicular artery (Adamkiewicz´s)*
→ Great radicular artery

Great wing of sphenoid bone
Skeleton 1
TA Latin: *Ala major ossis sphenoidalis*
TA English: *Greater wing of sphenoid*

GSE (general somato-effrent series)
General CNS
→ CSA (common somato-afferent series)

GSA (general somato-afferent series)
General CNS
→ CSA (common somato-afferent series)

Gustatory nucleus
The upper part of the solitary nucleus receives gustatory afferentes from facial, glossopharyngeal and vagus nerves. This part is therefore called the gustatory nucleus.

GVA (general viscero-afferent series)
General CNS
→ CSA (common somato-afferent series)

GVE (general viscero-efferent series)
General CNS
→ CSA (common somato-afferent series)

Gyrus rectus
Telencephalon 2
TA Latin: *Gyrus rectus*
TA English: *Straight rectus*

A long gyrus situated on the basal surface of the frontal lobe and running parallel to the olfactory tract and terminating, towards the brain, in the parolfactory sulcus.

H

Habenula

Diencephalon 1

TA Latin: *Habenula*

TA English: *Habenula*

The habenula is also called the epiphyseal stalk and is composed of three elements: the two habenular nuclei and habenular commissure, interconnecting the two nuclei. These three elements form a triangle which is also called the habenular trigone.

Little is known about the functions of the habenula. It can be viewed as being part of the limbic system.

Habenular commissure

Diencephalon 1

TA Latin: *Commissura habenularum*

TA English: *Habenular commissure*

The habenular nuclei of both sides are connected via fibers which pass in the wall of the third ventricle, at the base of the pineal body. These fibers form the habenular commissure.

Habenular ganglion

Nerves

TA Latin: *Nucl. habenularis*

TA English: *Habenular nucleus*

→ Habenular nucleus

Habenular nucleus

Diencephalon 1

TA Latin: *Nucl. habenularis*

TA English: *Habenular nuclei*

Together with the pineal body, the habenular nuclei form the epithalamus.

Situated above the superior colliculus, a distinction is made between the lateral, large-celled loosely packed lateral habenullar nucleus and the medial small-celled densely packed medial habenular nucleus. Efferents pass in the habenulointerpeduncular tract, and afferents come via the medullary stria. The habenula can be viewed as being part of the limbic system.

Habenular trigone

Diencephalon 1

TA Latin: *Trigonum habenulare*

TA English: *Habenular trigone*

At the transition to the thalamus are scattered the habenular nuclei, forming the epiphyseal stalk. The nuclei are scattered such that they form a triangle, called the habenular trigone.

Habenulo-interpeduncular tract (Meynert)

Diencephalon 2

TA Latin: *Tractus habenulointerpeduncularis*

TA English: *Habenulo-interpeduncular tract*

Subcortical efferents of the habenular nuclei leave this nuclear region via the habenulo-interpeduncular tract, also called fasciculus retroflexus of Meynert.

Hemianopsia

Dysfunction of the artery of the angular gyrus produces a combination of aphasia, alexia and hemianopsia.

Dysfunction of the supramarginal artery leads to hypoperfusion of the optic radiation, thus causing hemianopsia.

Damage to the visual cortex of one hemisphere leads to anything from disruption of fields of vision (scotoma), directly correlated with the extent of damage, to homonymous hemianopsia (semi-blindness with disruption of one eye field).

If both visual cortices are affected, cortical blindness results. Eye reflexes such as pupillary reflex are preserved, but the cortex-related accommodation reflex is lost.

Hemiazygos vein

Vessels 3

TA Latin: *V. hemiazygos*

TA English: *Hemi-azygos vein*

Drains venous blood from some thoracic organs, vertebral canal, vertebral column and posterior abdominal wall into the azygos vein.

Hemiballism

In the absence of inhibitory effect on the subthalamic nucleus, hyperkinesis of the contralateral proximal extremities' musculature presents. These typical involuntary movements are called ballism, and hemiballism in the case of unilateral expression.

Hemiparesis
Obstruction of the long central artery results in hemiparesis, paralytic symptoms in the tongue and facial musculature as well as in aphasia.

Heschl's junction
Telencephalon 3
TA Latin: *Gyri temporales transversi (Heschl)*
TA English: *Transverse temporal gyri (Heschl)*
→ Transverse temporal gyrus (Heschl)

Heubner's artery
Vessels
TA Latin: *A. Heubneri*
TA English: *Heubner's artery*
→ Long central artery (Heubner's)

Highest intercostal vein
Vessels
TA Latin: *V. intercostalis suprema*
TA English: *Supreme intercostal vein*
Veins coursing in the costal sulcus conducting blood from the back musculature, vertebral canal and vertebral column to the azygos vein and hemiazygos vein.

Hilum of olive
Cerebellum 2
TA Latin: *Hilum nuclei olivaris inf.*
TA English: *Hilum of inferior olivary nucleus*
The nucleus of the inferior olivary complex is the main nucleus in the olive and viewed in horizontal section, it features a horseshoe-shaped arrangement of its nuclei. The "open" part of the horseshoe is the point of entry/exit for afferent and efferent fibers and is called the hilum of olive.

Hippocampal formation
Telencephalon
TA Latin: *Hippocampus*
TA English: *Hippocampus*
→ Hippocampus

Hippocampal sulcus
Telencephalon 1
TA Latin: *Sulcus hippocampalis*

TA English: *Hippocampal sulcus*
Sulcus between dentate gyrus and subiculum.

Hippocampus
Telencephalon 3
TA Latin: *Hippocampus*
TA English: *Hippocampus*
The hippocampus is a C-shaped, simply-structured 3-layered allocortex in the medial wall of the telencephalon. It features 3 regions:

– hippocampus, precommissural part,
– hippocampus, supracommissural part,
– hippocampus, retrocommissural part.

The latter is divided longitudinally into the dentate fascia, Ammon's horn, subiculum. The hippocampus is involved in memory formation and the limbic system.
A vital component of the Papez neuronal circuit, the hippocampus is involved in memory formation. Lesions result in loss of the inability to transfer the contents from short-term memory to long-term memory (anterograde amnesia).

Hippocampus, precommissural part
Telencephalon 2
Situated in the subcallosal area, standing almost vertically this spindle-shaped collection of cells, which by virtue of its cell structure belongs to the hippocampus and joins the hippocampus, supracommissural part. Afferents come from the olfactory bulb and amygdaloid body. Efferents are unclear. Its function may be to couple olfactory stimuli with the limbic system.

Hippocampus, retrocommissural part
Telencephalon 3
The hippocampus, retrocommissural part, is the largest part of the hippocampus and consists of three regions:
– Subiculum,
– Ammon's horn
– dentate gyrus.

Hippocampus, supracommissural part
Telencephalon 2
The part of the hippocampus extending over the corpus callosum has three regions: the induseum griseum, containing cell bodies, and the two fiber strands of the longitudinal stria (medial and lateral). Afferents come from the

septum verum, the hippocampus, precommissural part, the medial thalamic nucleus and the hypothalamus. Efferents go to the corpus striatum and to the hippocampus, retrocommissural part.

Horizontal fissure (of cerebellum)

Cerebellum   3
TA Latin: *Fissura horizontalis (cerebelli)*
TA English: *Horizontal fissure (of cerebellum)*
The horizontal fissure is a large cerebellar groove running transversely across the semilunar lobule. It separates the superior semilunar lobule from the inferior semilunar lobule.

Hyperthermia

→ Medial preoptic nucleus

Hypoacusis

→ Inferior colliculus

Hypoglossal canal (incl. hypoglossal nerve)

Skeleton   1
TA Latin: *Canalis nervi hypoglossi*
(cum N. hypoglossus)
TA English: *Hypoglossal canal*
(incl. hypoglossal nerve)
Bony canal via which the hypoglossal nerve (XII) exits from the skull.

Hypoglossal nerve (XII)

Nerves   3
TA Latin: *N. hypoglossus (N.XII)*
TA English: *Hypoglossal nerve (XII)*
Hypoglossal nerve (XII) is a purely somatomotor nerve and innervates the tongue muscles. It has important functions in speaking, drinking, eating and swallowing.
Skull: hypoglossal canal.

Hypophysis

Diencephalon   3
TA Latin: *Hypophysis*
TA English: *Pituitary gland*
The hypophysis consists of the hypothalamic neurohypophysis (posterior lobe of the hypophysis) and the adenohypophysis (anterior lobe of the hypophysis) arising on the pharyngeal roof (Rathke's pouch). The hypophysis is the "hormone center" of the body and has close connections with the hypothalamus. With its

effector hormones and glandotropic hormones, it controls the autonomic processes and hormone glands of the body.

Hyposmia, anosmia

If the olfactory nerve (I) is damaged, e.g. as a sequel of a base of skull injury, depending on the extent of injury, hyposmia or anosmia can ensue. Pungent substances such as ammonia can still be smelt as the nasal mucosa are stimulated, stimulating in turn the trigeminal nerve.

Hypothalamic area, hypothalamic branch

Diencephalon   1
TA Latin: *A. communicans post.,*
R. hypothalamicus
TA English: *Posterior communicating artery,*
hypothalamic branch
Inconstant later branch of the posterior communicating artery.

Hypothalamic branch of the posterior communicating artery

Vessels
TA Latin: *R. hypothalamicus a. Communicans post.*
TA English: *Hypothalamic branch of the posterior communicating artery*
→ Posterior communicating artery, hypothalamic branch

Hypothalamic sulcus

Mesencephalon   1
TA Latin: *Sulcus hypothalamicus*
TA English: *Hypothalamic sulcus*
The hypothalamic sulcus is the name for the transition from thalamus to hypothalamus.

Hypothalamospinal fibers

Diencephalon   1
TA Latin: *Fibrae hypothalamospinales*
TA English: *Hypothalamospinal fibres*
Fibers of the posterior hypothalamic nucleus which pass without synapsing into the spinal cord. Their significance is unclear.

Hypothalamus

Diencephalon   3
TA Latin: *Hypothalamus*
TA English: *Hypothalamus*

The hypothalamus is the control center for basic autonomic processes (breathing, circulation, temperature, nutrients, fluid balance, sexual function).

Endocrine organs and autonomic nervous system are under its humoral and nerve control. The nuclei are subdivided into three groups (anterior, middle and posterior). The infundibulum and posterior lobe of the hypophysis are evaginations of the hypothalamus.

Hypothermia

$\rightarrow$ Medial preoptic nucleus

I

Indusium griseum

Telencephalon   1

TA Latein: *Indusium griseum*

Ta English: *Indusium griseum*

An element of the hippocampus, supracommissural part. Runs on the roof of the corpus callosum from the hippocampus, retrocommissural part, frontally to the hippocampus, precommissural part.

Inferior anastomotic vein (Labbé)

Vessels   2

TA Latin: *V. anastomotica inf. (Labbé)*

TA English: *Inferior anastomotic vein (Labbé)*

The inferior anastomotic vein connects the superficial middle cerebral vein with the transverse sinus.

The superior and inferior anastomotic veins form an anastomosis between the superior sagittal sinus and the transverse sinus.

Inferior artery of vermis

Vessels   1

Lateral branch of the superior cerebellar artery, medial branch.

Courses via the anterior lobe and along the cerebellar vermis, supplying the cortical regions of the cerebellum.

Inferior cerebellar peduncle

Cerebellum   3

TA Latin: *Pedunculus cerebellaris inf.*

TA English: *Inferior cerebellar peduncle*

Afferents: olivocerebellar tract (from olive), posterior spinocerebellar tract (from spinal cord, trunk) vestibulocochlear tract (directly and via vestibular nuclei), cuneocerebellar tract (from cuneate nucleus, neck), trigeminocerebellar tract (face). Efferents to vestibular nuclei: cerebellovestibular tract (directly from cerebral cortex), uncinate fasciculus (from fastigial nucleus).

Inferior cerebral veins

Vessels   2

TA Latin: *Vv. inf. cerebri*

TA English: *Inferior cerebral veins*

Collective designation for the superficial veins of the occipital lobe and temporal lobe, which collect venous blood from the cerebral cortex and flow directly into the transverse sinus.

Their counterpart is the superior cerebral veins conveying their blood content into the superior sagittal sinus.

Inferior choroid vein

Vessels   2

TA Latin: *V. choroidea inf.*

TA English: *Inferior choroid vein*

The inferior choroid vein carries venous blood from the choroid plexus of the lateral ventricle and hippocampus to the basal vein.

Inferior collicular branch

Vessels

→ Superior cerebellar artery, mesencephalic branch

Inferior colliculus

Mesencephalon   3

TA Latin: *Colliculus inf.*

TA English: *Inferior colliculus*

The lower of the two pairs of hills of the quadrigeminal plate (= tectum of mesencephalon) is part of the auditory tract and contains the central nucleus of the inferior colliculus. The inferior colliculus is part of the auditory tract and receives afferents via the lateral lemniscus from the cochlear nuclei and the nuclei of superior olive. Efferents pass via the brachium of inferior colliculus to the medial and lateral geniculate bodies as well as via the tectospinal tract to the spinal cord.

Dysfunction of an inferior colliculus induces hypoacusis, but not complete loss (since the auditory for each ear pathway runs bilaterally).

Inferior colliculus, lateral zone

Mesencephalon   1

Nuclear band running lateral to the central nucleus of the inferior colliculus.
Afferents come from the surrounding nuclei and the primary auditory cortex.
Efferents project to the contralateral cochlear nucleus and superior colliculus.

Inferior frontal gyrus
Telencephalon 3
TA Latin: *Gyrus front. inf.*
TA English: *Inferior frontal gyrus*
The inferior frontal gyrus comprises the following:
– inferior frontal gyrus, orbital part
– inferior frontal gyrus, triangular part
– inferior frontal gyrus, opercular part

In the areas of the frontal gyrus close to the precentral gyrus is situated the premotor cortex, which plays an important role in planning effector voluntary movements and has close interaction with the cerebellum, thalamic nuclei and basal ganglia.
In the inferior frontal gyrus, opercular part, lies the motor speech center (Broca). Here speech is planned but not executed.
Damage to the inferior frontal gyrus causes motor aphasia. Comprehension of spoken and written language is preserved, with mistakes occurring only on generating one's own language, whose severity correlates with the extent of damage and can range from impaired word-finding ability through agrammatism to complete loss of language.

Inferior frontal sulcus
Telencephalon 3
TA Latin: *Sulcus front. inf.*
TA English: *Inferior frontal sulcus*
The inferior frontal sulcus separates the medial frontal gyrus from the inferior frontal gyrus lying beneath it.

Inferior hemisphere vein
Vessels 2
TA Latin: *V. inf. cerebelli*
TA English: *Inferior vein of cerebellar hemisphere*
Superficial veins drain blood from the cerebellar cortex and carry it into the tentorial sinus (con-

fluence of the infratentorial veins). The latter flows into the confluence of the sinuses.

Inferior hypophyseal artery
Vessels 1
TA Latin: *A. hypophysialis inf.*
TA English: *Inferior hypophysial artery*
Arises from the internal carotid artery, cavernous part, and supplies the posterior lobe of the hypophysis.

Inferior intervertebral vein
Vessels 2
TA Latin: *V. intervertebralis*
TA English: *Intervertebral vein*
The intervertebral veins unite the hemiazygos vein flowing outside the vertebral canal with the internal vertebral venous plexus situated in the epidural cavity.

Inferior longitudinal fasciculus
Pathways 1
TA Latin: *Fasciculus longitudinalis inf.*
TA English: *Inferior longitudinal fasciculus*
Belongs to the association fibers and connects the temporal lobe and occipital lobe.

Inferior medullary velum
Cerebellum 3
TA Latin: *Velum medullare inf.*
TA English: *Inferior medullary velum*
The paired inferior medullary velum connects the cerebellum with the choroid tela of the fourth ventricle and is thus present on the roof of the fourth ventricle.

Inferior oblique muscle
Eye 2
TA Latin: *M. obliquus inf.*
TA English: *Inferior oblique muscle*
6 external eye muscles are distinguished. The following muscles rotate the eyeball:
– medial rectus muscle,
– lateral rectus muscle,
– inferior rectus muscle,
– superior rectus muscle,
– inferior oblique muscle,
– superior oblique muscle.
The function of the individual muscles is complex and depends in part on the respective start-

ing position of the eyeball. For more information, please refer to physiology textbooks.

Inferior occipito-frontal fasciculus

Telencephalon 2
TA Latin: *Fasciculus occipitofrontalis inf.*
TA English: *Inferior occipitofrontal fasciculus*
Association fibers connecting basal areas of the frontal lobe with the temporal and occipital lobes.

Inferior olivary complex

TA Latin: *Nucl. olivaris inf.*
TA English: *Inferior olivary nucleus*
→ Inferior olive

Inferior olive

Myelencephalon 3
TA Latin: *Oliva inf.*
TA English: *Inferior olive*
A large nucleus directly beside the pyramid, on the lower margin of the pons. Part of the motor system. Afferents come from the red nucleus and, as collaterals, from the pyramidal tract, from the precentral gyrus (motor cortex).
Efferents course as the olivocerebellar tract to the contralateral cerebellum, thus participating in the feedback loops between cerebellum and cortex regulating movement coordination.

Inferior ophthalmic vein

Vessels 1
TA Latin: *V. ophthalmica inf.*
TA English: *Inferior ophthalmic vein*
Belongs to the group of the orbital veins.
Collects venous blood from the palpebra inferior, lacrimal glands and eye muscles and carries it into the cavernous sinus.

Inferior parietal lobule

Telencephalon 3
TA Latin: *Lobulus parietalis inf.*
TA English: *Inferior parietal lobule*
In the direction of the occipital pole, the inferior and superior lobules unite at the postcentral gyrus.
Analogous to the secondary motor cortex there is also a secondary sensory cortex for the somatosensory control; this is believed to

stretch across both lobules and to be responsible for analysis, recognition and assessment of tactile information.
While dysfunctions of this area do not undermine tactile perception, they do subvert the accompanying recognition, judgmental and associative processes. The ensuing condition is called tactile agnosia.

Inferior petrosal sinus

Vessels 2
TA Latin: *Sinus petrosus inf.*
TA English: *Inferior petrosal sinus*
The inferior petrosal sinus lies on the petrous bone at the level of the exit points of the vagus nerve, accessory nerve and glossopharyngeal nerve.
It receives the labyrinth vein, veins of the cerebral pons, myelencephalon as well as the cerebellum.
It transports the blood through the base of the skull into the extracranially located superior bulb of the jugular vein.

Inferior petrosal vein (inconstant)

Vessels 1
In some cases a vein is formed which flows into the superior bulb of the jugular vein, close to the petrosal sinus.

Inferior posterior ventral nucleus

TA Latin: *Nucl. vent. post. inf.*
TA English: *Ventral posterior inferior nucleus of thalamus*
The ventral posterior (VP) nucleus is involved in somatosensory integration. Its medial segments (VPM) receive the somatotopically arranged trigeminothalamic projections, whereas the lateral segment (VPL) receives the spinothalamic projections. Epicritic and protopathic signals of the contralateral body half are processed. Projections go to the postcentral gyrus and surrounding areas.

Inferior rectus muscle

Eye 2
TA Latin: *M. rectus inf.*
TA English: *Inferior rectus muscle*
→ Inferior oblique muscle

Inferior sagittal sinus

Vessels   3

TA Latin: *Sinus sagittalis inf.*

TA English: *Inferior sagittal sinus*

The inferior sagittal sinus runs in the free margin of the falx cerebri, thus on the roof of the corpus callosum.

It collects the venous blood of the falx cerebri, corpus callosum, as well as of parallel cingulate gyrus and transports it to the straight sinus, which transports it further to the confluence of the sinuses.

Inferior semilunar lobule

Cerebellum   2

TA Latin: *Lobulus semilunaris inf.*

TA English: *Inferior semilunar lobule*

The inferior semilunar lobule belongs to the posterior lobe and is part of the cerebellar hemispheres. Apart from the areas in proximity to the vermis (intermediate part), the hemispheres belong to the phylogenetically young neocerebellum and receive their afferents via the mossy fibers of the pontocerebellar tract from the pontine nuclei. All hemisphere segments are hence also assigned to the pontocerebellum.

Inferior temporal gyrus

Telencephalon   3

TA Latin: *Gyrus temporalis inf*

TA English: *Inferior temporal gyrus*

The inferior temporal gyrus lies on the lower edge of the parietal lobe, passing on the basal side to the occipitotemporal sulcus.

Inferior temporal sulcus

Telencephalon   3

TA Latin: *Sulcus temporalis inf.*

TA English: *Inferior temporal sulcus*

The inferior temporal sulcus separates the medial temporal gyrus from the inferior temporal gyrus.

Inferior thalamostriate vein

Vessels   1

TA Latin: *V. thalamostriata inf.*

TA English: *Inferior thalamostriate vein*

→ Superior thalamostriate vein

Inferior vein of vermis

Vessels   2

TA Latin: *V. inf. vermis*

TA English: *Inferior vein of vermis*

The inferior vein of vermis runs along the lower portion of the vermis cerebelli, conveying its blood content either directly or via a detour into the straight sinus.

Inferior veins of cerebellar hemisphere

Vessels

TA Latin: *Vv. inf. cerebelli*

TA English: *Inferior veins of cerebellar hemisphere*

→ Inferior hemisphere vein

Inferior ventricular vein

Vessels   1

TA Latin: *V. ventricularis inf.*

TA English: *Inferior ventricular vein*

The inferior ventricular vein courses through the choroid fissure and carries venous blood from the deep region of the parietal lobe into the basal vein.

Inferior vestibular nucleus

Pons   2

TA Latin: *Nucl. vestibularis inf.*

TA English: *Inferior vestibular nucleus*

→ Vestibular nuclei (medial, superior , inferior)

Infraorbital artery

Vessels   2

TA Latin: *A. infraorbitalis*

TA English: *Infra-orbital artery*

The infraorbital artery arises from the maxillary artery and supplies the maxillae, gums, front teeth, maxillary sinus and orbita.

Infraorbital canal

Skeleton   1

TA Latin: *Canalis infraorbitalis*

TA English: *Infra-orbital canal*

Bony canal in floor of orbita.

Infraorbital foramen

Meninges & Cisterns   1

TA Latin: *Foramen infraorbitalis*

TA English: *Infra-orbital foramen*

Exit of the similarly named canal on the lower margin of the orbita. Passage of the infraorbital artery and infraorbital nerve.

Infraorbital vein

Vessels 1

Belongs to the group of the orbital veins.
A vein running along the lower wall of the orbita and forming an anastomosis between the pterygoid plexus and the facial vein.

Infundibular nucleus

Diencephalon 2
TA Latin: *Nucl. arcuatus*
TA English: *Arcuate nucleus*

This nucleus, also called arcuate nucleus, lies in the intermediate group of the medial zone of the hypothalamus in the tuber cinerum, directly at the attachment of the infundibulum. It contains neurons producing release and release-inhibiting factors, which regulate hormone secretion in the anterior lobe of the hypophysis. The axons of these neurons course in the tuberoinfundibular tract to the median eminence, where they anchor at blood vessels.

Infundibular recess

Meninges & Cisterns 2
TA Latin: *Recessus infundibuli*
TA English: *Infundibular recess*

A narrow extension of the third ventricle penetrates into the hypophyseal stalk (infundibulum).

Infundibulum

Diencephalon 2
TA Latin: *Infundibulum*
TA English: *Infundibulum*

The infundibulum is the neurohypophyseal part of the hypophyseal stalk. It emerges from the tuber cinerum and, in its upper segment, it surrounds an evagination of the third ventricle, the infundibular recess. In the infundibulum run primarily the axons of the supraoptic nucleus and paraventricular nucleus, which release their hormones ADH and oxytocin into the blood.

Inion

Skeleton 1
TA Latin: *Inion*
TA English: *Inion*

Inion is the summit of the occipital protuberance.

Insula (of Reil)

Telencephalon 3
TA Latin: *Insula*
TA English: *Insula*

Phylogenetically, the insula is very old and in the course of brain evolution, it has been overlain by other lobes of the brain (operculum = lid). Its cortical region (cortex insula) is separated by the circular sulcus of the insula from the surrounding lobes.

The function of this region is not really known, but viscerosensory and visceromotor functions are supposed.

Insular cortex

Telencephalon 3
→ Insula (of Reil)

Insular pole

Telencephalon 2

The rostralmost part of the insula. Joins the rostral perforated substance.

Insular veins

Vessels 1
TA Latin: *Vv. insulares*
TA English: *Insular veins*

Superficial veins at the temporal pole. They collect the blood from the surrounding area of the brain tissue and transport it into the anterior cerebral veins.

Intention tremor

Lesions in the red nucleus detract from the basic tone of body musculature and produce intention tremor and chorea-like movements.

Intercalate nucleus

Myelencephalon 1
TA Latin: *Nucl. intercalatus*
TA English: *Intercalated nucleus*

The intercalate nucleus belongs with the prepositus hypoglossal nucleus of the perihypoglossal nuclei and sends, as is typical of this nuclear group, direct efferents to the nuclei of eye muscles.

Intercollicular nucleus

Mesencephalon   1

Nuclear region between the colliculi of the quadrigeminal plate.

Intercollicular zone

→ Intercollicular nucleus

Intercostal nerve
(ventral branch of thoracic nerve)

Nerves   2

TA Latin: *N. intercostalis (R. vent. n. thoracici)*
TA English: *Intercostal nerve (ventral ramus of thoracic nerve)*

Whereas the ventral branches of the spinal nerve normally unite to form a plexus, this is not the case in the thoracic cord. The ventral branches course as intercostal nerves, segmentally separated, to the body periphery.

Intercrural cistern

Meninges & Cisterns
TA Latin: *Cisterna interpeduncularis*
TA English: *Interpeduncular cistern*

→ Interpeduncular cistern

Intermediary nerve

Nerves   2

TA Latin: *N. intermedius*
TA English: *Intermediate nerve*

The facial nerve conducts three qualities:

1) motor efferents for innervating the mimetic muscles. Nucleus: nucleus of the facial nerve.
2) visceromotor control: parasympathetic innervation of salivary and lacrimal glands. Nucleus: salivatory nuclei.
3) Somatosensory control: sensory innervation of the tongue (anterior 2/3) and of the external ear. Nucleus: solitary nucleus.

Qualities (2) and (3) are mediated by the intermediate maxillary nerve (V2) which is a permanent part of the facial nerve.

Intermediate gray substance

TA Latin: *Subst. intermedia centralis*

TA English: *Central intermediate substance*
→ Intermediate substance

Intermediate internal frontal artery

Vessels
TA Latin: *A. callosomarginalis, R. Front. intermediomed.*
TA English: *Callosomarginal artery, intermediomedial frontal branch*

→ Callosomarginal artery, intermediomedial frontal branch

Intermediate substance

Medulla spinalis   3

TA Latin: *Subst. intermedia*
TA English: *Intermediate substance*

The gray matter of the spinal cord between posterior horn and anterior horn.

Here the afferents of the touch receptors synapse, as do the viscero-afferents.

Afferents from the joints terminate here in rather ventromedial sections.

Intermediate temporal artery

Vessels   1

TA Latin: *A. cerebri media, R. Temporalis medius*
TA English: *A. cerebri media, middle temporal branch*

Arises from the middle cerebral artery, posterior trunk. The middle cerebral artery emerges from the internal carotid artery.

Supplies the middle portion of the temporal lobe.

Intermediate trunk

Vessels

→ Middle cerebral artery, parietal trunk

Intermediolateral nucleus

Medulla spinalis   2

TA Latin: *Nucl. intermediolat.*
TA English: *Intermediolateral nucleus*

Also called intermediolateral substance. Part of the lateral horn in the thoracic cord with sympathetic neurons for vasomotor activities.

Intermediolateral nucleus, gelatinous substance
Medulla spinalis 1
Efferents of the paraventricular nucleus terminate in the intermediolateral nucleus and gelatinous substance (of Roland) of the spinal cord.

Intermediomedial frontal branch
Vessels
TA Latin: *R. front. intermediomed.*
TA English: *Intermediomedial frontal branch*
→ Callosomarginal artery,
intermediomedial frontal branch

Internal arcuate fibers
Mesencephalon 2
TA Latin: *Fibrae arcuatae internae*
TA English: *Internal arcuate fibres*
The efferents of the gracile and cuneate nuclei cross the mesencephalon, ascending in the contralateral medial lemniscus.

Internal capsule
Telencephalon 3
TA Latin: *Capsula interna*
TA English: *Internal capsule*
Virtually all ascending and descending cortical pathways pass through the internal capsule. The capsule is v-shaped in horizontal sections and is composed of three regions:
– anterior limb of internal capsule,
– genu of internal capsule,
– posterior limb of internal capsule.
Depending on the position vis-à-vis the lentiform nucleus, two other portions can also be distinguished in the posterior limb:
– retrolenticular part of internal capsule,
– sublenticular part of internal capsule.

Internal capsule branch
Vessels
TA Latin: *A. choroidea ant.,*
Rr. Capsulae
internae
TA English: *Anterior choroidal artery,*
branches to internal capsule
→ Anterior choroid artery, internal capsular branches

Internal carotid artery
Vessels 3
TA Latin: *A. carotis interna*
TA English: *Internal carotid artery*
Emerges with the external carotid artery from the division of the common carotid artery. Passes via the lateropharyngeal space to the base of the skull, courses through the carotid canal and the lacerate foramen into the cavernous sinus, and from there through the dura mater into the middle cranial fossa. Accordingly, four segments are distinguished: cervical part, petrous part, cavernous part and cerebral part.

Internal carotid artery, cavernous part (carotid siphon)
Vessels 3
TA Latin: *A. carotis interna, pars cavernosa*
TA English: *Internal carotid artery, cavernous part (carotid siphon)*
This section of the internal carotid artery is S-shaped and comprises five segments, of which the lower siphon limb, the siphon genu and the upper siphon limb together form the carotid siphon. The other two segments are the so-called ganglion segment and posterior cavernous curvature. The S-shape of this segment may be implicated in compensating major pressure fluctuations.

Internal carotid artery, cerebral part
Vessels 2
TA Latin: *A. carotis interna, pars cerebralis*
TA English: *Internal carotid artery, cerebral part*
This final section of the internal carotid artery courses the chiasmatic cistern and divides at the base of the skull into two terminal branches, the anterior cerebral artery and the middle cerebral artery. Together with the posterior cerebral artery, which emerges from the basilar artery, these two represent the three major arteries of each hemisphere.

Internal carotid artery, cervical part
Vessels 2
TA Latin: *A. carotis interna, pars cervicalis*
TA English: *Internal carotid artery, cervical part*

This segment of the internal carotid artery features the carotid sinus with its sensors, which play an important role in blood pressure regulation and homeostasis. After a projecting vascular loop, which offsets movements of the skull, the vertebral artery, transverse part enters the carotid canal together with the vagus nerve and the internal jugular vein, and is then called the petrous part.

Internal carotid artery, petrous part

Vessels   2
TA Latin: *A. carotis interna, pars petrosa*
TA English: *Internal carotid artery, petrous part*
In the carotid canal, the internal carotid artery passes through the petrous part (petrous bone) of the temporal bone. Within the canal, it courses close to the cochlea and past the trigeminal ganglion. On exiting from the canal, the cavernous part of the internal carotid artery begins.

Internal cerebral vein

Vessels   2
TA Latin: *V. int. cerebri*
TA English: *Internal cerebral vein*
The thalamostriate vein and superior choroid vein enter the internal cerebral vein which flows into the great cerebral vein at the level of the pineal gland.

Internal frontal artery

Vessels
TA Latin: *A. Callosomarginalis, R. Front. anteromed.*
TA English: *Callosomarginal artery, anteromedial frontal branch*
→ Callosomarginal artery, anteromedial frontal branch

Internal iliac vein

Vessels   3
TA Latin: *V. iliaca int.*
TA English: *Internal iliac vein*
Drains venous blood from the pelvic intestines and buttocks into the common iliac vein.

Internal jugular vein

Vessels   3
TA Latin: *V. jugularis interna*
TA English: *Internal jugular vein*
Large jugular vein, running at a deep level, which drains venous blood from the sigmoid sinus, face, pharynx, brain, tongue, larynx and thyroid gland into the brachiocephalic vein, which in turn flows into the superior vena cava.

Internal maxillary artery

Vessels
TA Latin: *A. maxillaris*
TA English: *Maxillary artery*
→ Maxillary artery

Internal medullary lamina of the thalamus

Diencephalon   2
TA Latin: *Lamina medullaris med.*
TA English: *Internal medullary lamina of thalamus*
This internal medullary layer subdivides the thalamus into three large thalamic nuclear groups:
– medial thalamic nucleus,
– ventral thalamic nucleus and
– lateral thalamic nucleus.

Internal occipital protuberance

Skeleton   2
TA Latin: *Protuberantia occipitalis interna*
TA English: *Internal occipital protuberance*
Strong bony elevation on the inside of the occipital bone. From here the falx cerebri stretches to the christa galli.

Internal vertebral venous plexus

Vessels   3
TA Latin: *Plexus venosus vertebralis internus*
TA English: *Internal vertebral venous plexus*
Greatly anastomosed anterior internal vertebral venous plexus and posterior internal vertebral venous plexus are venous plexuses in the vertebral canal. They collect blood from the vertebral canal and transport it via the venous plexuses of the intervertebral canals to the veins outside the vertebral canal.
Via the basivertebral veins they are connected

with the posterior external vertebral venous plexus.

Interpeduncular cistern

Meninges & Cisterns 2

TA Latin: *Cisterna interpeduncularis*
TA English: *Interpeduncular cistern*
The interpeduncular cistern is formed by the two cerebral peduncles. In the interpeduncular cistern courses the precommunical part of the posterior cerebral artery.

Interpeduncular fossa

Meninges & Cisterns 2

TA Latin: *Fossa interpeduncularis*
TA English: *Interpeduncular fossa*
Above the pons are encountered two large, v-shaped fiber bundles running parallel to each other and containing efferents which descend in the direction of the brainstem and spinal cord. These two strands are called the cerebral peduncles and the groove between them is the interpeduncular fossa.

Interpeduncular nucleus

Mesencephalon 1

TA Latin: *Nucl. interpeduncularis*
TA English: *Interpeduncular nucleus*
A nucleus belonging to the tegmentum of mesencephalon in the basal part of the interpeduncular fossa.
Afferents come from the thalamic pulvinar and the pretectal region.
Efferents go to the dorsal tegmental nucleus (Gudden).
Appears to be implicated in autonomic regulatory mechanisms.

Interpeduncular vein

Vessels 1

TA Latin: *V. interpeduncularis*
TA English: *Interpeduncular vein*
The small interpeduncular veins at the cerebral peduncles drain the hypothalamus, subthalamus and optic chiasm.

Interpositus nucleus
(emboliform nucleus + globose nucleus)

Cerebellum 1

TA Latin: *Nucl. Interpositus*
(Nucl. emboliformis + globosus)
TA English: *Interpositus nucleus*
(emboliform nucleus + globose nucleus)
Emboliform nucleus as well as globose nucleus receive their afferences from the Purkinje's corpuscles of the cerebellar hemisphere, intermediate pars. Therefore they are also summed up as interposite nucleus. They are also called "intermediary nuclei".

Interstitial nucleus (Cajal)

Mesencephalon 1

TA Latin: *Nucl. interstitialis (Cajal)*
TA English: *Interstitial nucleus (Cajal)*
Darkschewitsch´s nucleus and the interstitial nucleus (Cajal) are two groups of small cells within the reticular formation. Afferents arise from the ipsilateral corpus striatum and globus pallidus, vestibular nuclei as well as the contralateral cerebellum. Efferents pass across the interstitiospinal tract into the spinal cord. This nucleus is part of the motor system (oculomotor) and shares responsibility for regulation of muscle tone.

Interstitial nucleus of the stria terminalis

Diencephalon 2

The stria terminalis, most important amygdaloid efferent, divides into three bundles at the anterior commissure. One bundle, the postcommissural stria terminalis, terminates here.
Afferents: insula, subiculum, amygdaloid body, hypothalamus, myelencephalon.
Efferents: amygdaloid body, hypothalamus, thalamus, brainstem.
Function: regulation of cardiovascular and respiratory components as well as of male sexual behavior.

Interstitial rostral nucleus of the medial longitudinal fasciculus

Mesencephalon 1

TA Latin: *Nucl. interstitialis fasciculi longitudinalis med.*
TA English: *Interstitial nucleus of the medial longitudinal fasciculus*
Like the adjacent interstitial nucleus (Cajal), this nucleus is also involved in control of extraocular

muscles. Afferents come from the vestibular nuclei.

Interstitiospinal tract
Pathways   2
TA Latin: *Tractus interstitiospinalis*
TA English: *Interstitiospinal tract*
Projections of the interstitial nucleus (Cajal) to the spinal cord. Runs in the medial longitudinal fasciculus.

Interthalamic adhesion
Mesencephalon   1
TA Latin: *Adhesio interthalamica*
TA English: *Interthalamic adhesion*
A bridge of gray matter between both thalami.

Interventricular foramen (of Monro)
Meninges & Cisterns   3
TA Latin: *Foramen interventriculare (Monroi)*
TA English: *Interventricular foramen (of Monro)*
Connection between the lateral ventricle and third ventricle.

Intervertebral foramen
Meninges & Cisterns   2
TA Latin: *Foramen intervertebrale*
TA English: *Intervertebral foramen*
The intervertebral foramen is the point of passage for the spinal nerves.
By means of connective tissue strands, the dural sack of spinal ganglia and spinal roots is suspended here.

Intralaminar nucleus
Diencephalon
TA Latin: *Nucl. intralaminaris thalami*
TA English: *Intralaminar nucleus of thalamus*
→ Intralaminar thalamic nuclei

Intralaminar thalamic nuclei
Diencephalon   2
TA Latin: *Nuclei intralaminares thalami*
TA English: *Intralaminar nuclei of thalamus*
The intralaminar nuclei lie in the internal medullary lamina of the thalamus and are characterized by their double projections, with one going to the cerebral cortex and one to the corpus striatum. A distinction is made between
(1) rostral group: lateral central nucleus, paracentral nucleus, medial central nucleus.
(2) caudal group: centromedian nucleus, parafascicular nucleus.
Afferents from the globus pallidus, cerebellum, spino/trigeminothalamic tract.

Intralimbic gyrus
Telencephalon   2
Part of Ammon´s horn and hence a vital part of the hippocampus, limbic system and memory formation.

Intraparietal sulcus
Telencephalon   2
TA Latin: *Sulcus intraparietalis*
TA English: *Intraparietal sulcus*
Sulcus between the superior parietal lobule and the inferior parietal lobule.

Iris
Eye   2
TA Latin: *Iris*
TA English: *Iris*
Part of the eye. Used for fast and precise adaptation (bright-dark adaptation).

Isocortex
Telencephalon   3
TA Latin: *Isocortex*
TA English: *Isocortex*
→ Neocortex

Isthmus of cingulate gyrus
Telencephalon   1
TA Latin: *Isthmus gyri cinguli*
TA English: *Isthmus of cingulate gyrus*
Narrowing of the cingulate gyrus at the splenium of the corpus callosum.

JKL

Jugular foramen

Meninges & Cisterns 3

TA Latin: *Foramen jugulare*

TA English: *Jugular foramen*

A large window in the base of the skull through which the internal jugular vein and various cranial nerves (IX, X, XI) pass.

Kölliker-Fuse nucleus

Diencephalon 2

TA Latin: *Nucl. Subparabrachialis (Kölliker-Fuse)*

TA English: *Subparabrachial nucleus (Kölliker-Fuse)*

The nucleus belongs to the parabrachial area and partially corresponds to the functionally defined pneumotactic center. The nucleus receives afferents from the caudal part of the solitary nucleus and the ventrolateral superficial reticular area.

Efferents ascend to the preoptic area and to the central amygdaloid nucleus.

The most important efferents go to the lower myelencephalon and the spinal cord.

Korsakov syndrome

Selective dysfunction of the mammillary body, medial nucleus results in Korsakoff syndrome (amnetic syndrome with anterograde and retrograde impaired memory, and diminished drive).

L-region of Holstege et al.

Mesencephalon 1

In the lateral tegmentum of pons there are 2 separate regions for regulating micturition. One region, the M-region of Holstege, is located in the dorsomedial portion of the tegmentum of pons. The other one lies in the ventrolateral tegmentum of pons and has been designated by Holstege as the L-region.

Labyrinthine artery

Vessels 1

TA Latin: *A. labyrinthi*

TA English: *Labyrinthine artery*

Arises from the anterior inferior cerebellar artery which for its part branches off from the basilar artery. It passes into the temporal bone and supplies the organ of equilibrium (semicircular canals).

Lacrimal artery

Vessels 2

TA Latin: *A. lacrimalis*

TA English: *Lacrimal artery*

Arises from the ophthalmic artery and supplies the lateral angle of eye, lateral eye muscles, conjunctiva of the eyeball and – as indicated by the name – lacrimal glands.

Lamina affixa

Diencephalon 2

TA Latin: *Lamina affixa*

TA English: *Lamina affixa*

A layer of epithelium growing on the surface of the thalamus and forming the floor of the lateral ventricle, central part, on whose medial margin is attached the choroid plexus of the lateral ventricle. The torn edge of this plexus is called the choroid tenia.

Lamina terminalis

Meninges & Cisterns 2

TA Latin: *Lamina terminalis*

TA English: *Lamina terminalis*

Anterior closure of the third ventricle.

In the lamina terminalis fibers from the septal verum (nucleus accumbens, diagonal band) ascend and pass via the corpus callosum to the posterior hippocampus.

Laminar tectum branch

Vessels

→ Superior cerebellar artery, mesencephalic branch

Lateral amygdaloid nucleus

Telencephalon

TA Latin: *Nucl. amygdalae lat.*

TA English: *Lateral amygdaloid nucleus*

→ Amygdaloid body

Lateral atrial vein

Vessels   1

A vein descending in the direction of the hippocampus and coursing along the side wall of the lateral ventricle.

Lateral central nucleus

Diencephalon   1

TA Latin: *Nucl. centralis lat.*
TA English: *Central lateral nucleus*

Belongs to the rostral group of intralaminar thalamic nuclei.

Receives afferents from the cerebellum, motor and parietal cortex as well as the spinothalamic tract.

Lateral column

Medulla spinalis   3

TA Latin: *Funiculus lat.*
TA English: *Lateral funiculus*

The white matter between the ventral root and dorsal root gives rise to the lateral column, containing:

1) anterolateral column with
   – anterolateral fasciculus
   – parts of the anterior spinocerebellar tract.
2) posterolateral column with
   – posterior spinocerebellar tract
   – parts of the anterior spinocerebellar tract
   – lateral pyramidal tract.

Lateral corticospinal tract

Pathways

TA Latin: *Tractus corticospinalis lat.*
TA English: *Lateral corticospinal tract*
→ Lateral pyramidal tract

Lateral cuneate nucleus

Myelencephalon
→ Medial cuneate nucleus

Lateral epidural vein

Vessels   2

Together, the lateral epidural vein and the middle epidural vein form the anterior internal vertebral venous plexus.

They drain the spinal cord and spinal meninges.

Lateral frontobasal artery

Vessels   2

TA Latin: *A. frontobasalis lat.*
TA English: *Lateral frontobasal artery*

Arises from the middle cerebral artery, frontal trunk. The middle cerebral artery for its part emerges from the internal carotid artery.

Via the inferior branch, it supplies parts of the frontal lobe, while the external branch supplies the medial and inferior frontal gyri.

Lateral funicle nucleus

Myelencephalon   1
→ Anterior funicle nucleus

Lateral geniculate body (LGB)

Diencephalon   3

TA Latin: *Corpus geniculatum lat. (CGL)*
TA English: *Lateral geniculate body (LGB)*

The LGB consists of the dorsal nucleus (dLGB) and ventral nucleus (vLGB). In humans the vLGB is merely a small cluster of cells, whereas the dLGB has several layers. Two magnocellular layers receive afferents from fast-conducting alpha ganglion cells of the retina, whereas the 4 parvocellular layers receive afferents from the smaller beta retinal ganglion cells. Both project via the optic radiation into lamina IV of the visual cortex, but to different layers.

Lateral geniculate body, magnocellular laminae

Diencephalon   2

The two magnocellular layers of the dorsal LGB of humans receive afferents from fast-conducting alpha cells of the retina (Y system) and project via the optic radiation to an intermediate sublayer of lamina IV of the visual cortex as well as to lamina I and to the border of Lamina V and VI.

Layer 1 receives fibers from the contralateral eye, and layer 2 from the ipsilateral.

Lateral geniculate body, parvocellular laminae

Diencephalon   2

The four parvocellular layers of the dorsal LGB of humans receive afferents from slow-conducting cells of the retina (X system) and project to the deep and superficial sublayers of lamina IV of the visual cortex.

Layers 4 and 6 receive fibers from the contralateral eye, and layer 3 and 5 from ipsilateral.

Lateral habenular nucleus

Diencephalon 2

TA Latin: *Nucl. habenularis lat.*

TA English: *Lateral habenular nucleus*

The lateral habenular nucleus belongs to the epithalamus and is composed of the large, loosely packed, polygonal cells.

Afferents primarily from the lateral hypothalamic area and the globus pallidus, but also from the diagonal band, substantia innominata and preoptic area.

Efferents go via the habenulointerpeduncular tract to the mesencephalon.

The functional significance of the nucleus remains unclear.

Lateral hypothalamic area

Diencephalon 2

TA Latin: *Area hypothalamica lat.*

TA English: *Lateral hypothalamic area*

The lateral zone of the hypothalamic gray matter cannot be clearly differentiated and is hence called the lateral hypothalamic area. The gray matter endowed with myriad links of its own receives manifold afferents and sends various efferents, also to motor centers. Due to its pronounced crosslinks, it is an important integration and computational center of the autonomic nervous system.

Lateral inferior artery of cerebellar hemisphere

Vessels

TA Latin: *A. inf. post. cerebelli, R. lat.*

TA English: *Posterior inferior cerebellar artery, lateral branch*

→ Posterior inferior cerebellar artery, lateral branch

Lateral lacunae

Vessels 3

TA Latin: *Lacunae lat.*

TA English: *Lateral lacunae*

Lateral evaginations of the superior sagittal sinus.

The diploic veins and meningeal vein enter this.

Lateral lemniscal nucleus

Mesencephalon 2

TA Latin: *Nucl. lemnisci lat.*

TA English: *Nucleus of lateral lemniscus*

A cellular column in the middle of the lateral lemniscus in which some of the fibers of the lateral lemniscus terminate. The nucleus has two parts:

- dorsolateral lemniscal nucleus: has projections to its counterpart on the contralateral side as well as to the contralateral inferior colliculus.
- Ventrolateral lemniscal nucleus.
  Both parts project directly to the ipsilateral medial geniculate body and to the ipsilateral inferior colliculus.

Lateral lemniscus

Mesencephalon 3

TA Latin: *Lemniscus lat.*

TA English: *Lateral lemniscus*

An element of the auditory tract. Contains crossed and uncrossed fibers from the cochlear nuclei and the nuclei of the superior olive, conveying these to the inferior colliculus and thus being an important component of the central auditory tract (second and third order neurons). From the inferior colliculus it passes further via the brachium of inferior colliculus to the lateral geniculate body, and from here via the auditory radiation to the primary auditory cortex.

Lateral mammillary nucleus

2

TA Latin: *Nucl. mammillaris lat.*

TA English: *Lateral nucleus of mammillary body*

→ Mammillary body, lateral nucleus

Lateral medullary lamina

Telencephalon 1

TA Latin: *Lamina medullaris lat.*

TA English: *Lateral medullary lamina*

Layer of white matter between the globus pallidus and putamen.

Lateral mesencephalic vein

Vessels   2
TA Latin: *V. mesencephalica lat.*
TA English: *Lateral mesencephalic vein*
The lateral mesencephalic vein is an anastomosis between the supra- and infratentorial system, by virtue of the fact that it connects the supratentorial basal vein with the infratentorial superior peduncular cerebellar vein.

Lateral nucleus of the superior olive

TA Latin: *Nucl. olivaris sup. lat.*
TA English: *Lateral superior olivary nucleus*
→ Nucleus of the superior lateral olive

Lateral occipital artery

Vessels   3
TA Latin: *A. occipitalis lat.*
TA English: *Lateral occipital artery*
Together with the middle occipital artery, it forms the terminal part of the posterior cerebral artery.
It supplies the basal segments of the occipital lobe as well as the posterior sections of the temporal lobe.

Lateral occipital artery, anterior temporal branches

Vessels   1
TA Latin: *A. occipitalis lat., Rr. temporales ant.*
TA English: *Lateral occipital artery, anterior temporal branches*
Lateral branches of the lateral occipital artery that embark on a varying course.

Lateral occipital artery, middle intermediate temporal branches

Vessels   1
TA Latin: *A. occipitalis lat., Rr. Temporales intermedii med.*
TA English: *Lateral occipital artery, middle intermediate temporal branches*
An inconstant lateral branch of the posterior cerebral artery, passing into the collateral sulcus, and which sometimes arises from the occipital artery.

Lateral occipital artery, posterior temporal branches

Vessels   1
TA Latin: *A. occipitalis lat., Rr. Temporales post.*
TA English: *Lateral occipital artery, posterior temporal branches*
→ Lateral occipital artery, anterior temporal branches

Lateral occipitotemporal gyrus

Telencephalon   3
TA Latin: *Gyrus occipitotemporalis lat.*
TA English: *Lateral occipitotemporal gyrus*
On the underside of the hemisphere, two well-developed gyri spread across the occipital lobe and temporal lobe. They are called:

– lateral occipitotemporal gyrus: runs parallel to the hippocampal gyrus, which lies on the other side of the collateral sulcus. This marks a transition zone between the allocortex and cerebral cortex.
– medial occipitotemporal gyrus: area 17, the striate cortex, is situated in the occipital portion, directly on the calcarine sulcus.

Lateral olfactory stria

Telencephalon   3
TA Latin: *Stria olfactoria lat.*
TA English: *Lateral olfactory stria*
At the brain end, the olfactory bulb divides into the lateral stria and the medial stria.
The lateral stria courses laterally, twisting sharply around the limen insula before entering the rostromedial part of the temporal lobe (ambiens gyrus, amygdaloid body).

Lateral orbitofrontal branch

Vessels
TA Latin: *A. frontobasalis lat.*
TA English: *Lateral frontobasal artery*
→ Lateral frontobasal artery

Lateral parabrachial nucleus

Diencephalon   2
TA Latin: *Nucl. parabrachialis lat.*
TA English: *Lateral parabrachial nucleus*
This nucleus plays an important role in processing gustatory signals. It receives major afferents

from the rostral gustatory segment of the solitary nucleus.

Its efferents go to the medial parabrachial nucleus, lateral hypothalamic area, substantia innominata and the medial hypothalamus.

This nucleus is also called the "pontine gustatory field".

Lateral paragigantocellular nucleus

Myelencephalon 1

TA Latin: *Nucl. paragigantocellularis lat.*

TA English: *Lateral paragigantocellular reticular nucleus*

Comprises the noradrenergic cell group A5 and dispatches efferents to the locus coeruleus.

Lateral pontine arteries

Vessels 2

TA Latin: *Aa. pontis, R. lat*

TA English: *Pontine arteries, lateral branch*

Small lateral arteries often are given off from the basilar artery into the surrounding tissue of the pons.

Some of these enter the pons more laterally, and some more medially.

Hence a distinction is made between lateral pontine arteries and medial pontine arteries.

Depending on their course, a further distinction can be made between inferolateral and superolateral branches.

Lateral posterior choroid artery

Vessels

TA Latin: *A. cerebri post.,*

R. choroideus post. lat.

TA English: *Posterior cerebral artery, posterior lateral choroidal branch*

→ Posterior cerebral artery, postcommunical part

Lateral preoptic area

Diencephalon

TA Latin: *Area hypothalamica lat.,*

Area praeoptica

TA English: *Lateral hypothalamic area, preoptic area*

→ Lateral preoptic nucleus

Lateral preoptic nucleus

Diencephalon 2

TA Latin: *Nucl. preopticus lat.*

TA English: *Lateral preoptic nucleus*

Although spatially and in respect of nomenclature closely related to the medial preoptic nucleus, this region appears to be part of a totally different set of functions, i.e. regulation of locomotion. Afferents come from the amygdaloid body and the majority of surrounding nuclear regions. Efferents go to the diagonal band, mammillary body, supramammillary region, tegmentum area and habenula.

Lateral pyramidal tract

Nerves 3

In the pyramidal decussation 70–90% of the fibers cross to the contralateral side forming the lateral pyramidal tract, descending in the lateral column of the spinal cord. The tract features a somatotopic arrangement (the lateralmost fibers pass to the lowest, the medialmost to the highest spinal segment). Its fibers terminate in the motoneurons (or interneurons of the motoneurons) of the distal extremity (forearm, hand) and play a major role in fine motor control.

Lateral recess branch of the fourth ventricle

Vessels 1

TA Latin: *A. inf. post. Cerebelli,*

R. Choroideus ventriculi quarti

TA English: *Posterior inferior cerebellar artery, choroidal branch to fourth ventricle*

A small, inconstant lateral branch of the posterior inferior cerebellar artery.

Courses to the choroid plexus of the fourth ventricle.

Lateral recess of the fourth ventricle

Meninges & Cisterns 2

TA Latin: *Recessus lat. ventriculi quarti*

TA English: *Lateral recess of fourth ventricle*

The extension of the fourth ventricle, which runs beneath the inferior and middle cerebellar peduncles, is called the lateral recess of the fourth ventricle.

Lateral recess vein of the fourth ventricle

Vessels 1

TA Latin: *V. recessus lat. ventriculi quarti*

TA English: *Vein of lateral recess of fourth ventricle*

A vein branching off from the pontomedullary vein to the lateral recess of the fourth ventricle. It drains the choroid plexus and flows into the inferior petrosal sinus.

Lateral reticular formation
Pons   2
This small-celled region of the reticular formation is limited to the pons and medulla. It has 6 different sections:
– ventrolateral superficial reticular area,
– parvocellular reticular area,
– lateral pontine area,
– noradrenergic cell groups A1–A7,
– adrenergic cell groups (C1, C2),
– cholinergic cell group (Ch1–Ch6).
These areas are involved in brainstem reflexes, cardiovascular, respiratory and gastrointestinal regulation and pain suppression.

Lateral reticular nucleus   2
TA Latin: *Nucl. reticularis lat.*
TA English: *Lateral reticular nucleus*
→ Lateral funicle nucleus

Lateral sacral vein
Vessels   2
TA Latin: *V. sacralis lat.*
TA English: *Lateral sacral vein*
Collects venous blood from the anterior sacrum and carries it into the internal iliac vein.

Lateral septal nucleus
Telencephalon   2
TA Latin: *Nucl. septalis lat.*
TA English: *Lateral septal nucleus*
Belongs to the septal nuclei. Receives its largest afferents from Ammon´s horn as well as from the subiculum of the hippocampus. Also from the preoptic area, hypothalamus, locus coeruleus, rahpe nuclei, Kölliker-Fuse nucleus and from the vagus complex. Efferents to other septal regions, preoptic area, thalamic nuclei of the midline as well as to the medial habenular nucleus.

Lateral sinus
Vessels   3
TA Latin: *Sinus transversus*
TA English: *Transverse sinus*

→ Transverse sinus

Lateral sulcus
Telencephalon   3
TA Latin: *Sulcus lat.*
TA English: *Lateral sulcus*
= Sylvian fissure.
Large and deep lateral sulcus, caudally contiguous with the temporal lobes and supporting the insula at its deep level.
The lateral sulcus has two important lateral branches: the posterior branch around which the supramarginal gyrus lies, as well as the ascending branch and anterior branch, around which the inferior frontal gyrus is grouped. The primary and secondary auditory cortices are grouped at the lateral sulcus.

Lateral sulcus, ascending branch
Telencephalon   3
TA Latin: *Sulcus lat., R. ascendens*
TA English: *Lateral sulcus, ascending ramus*
Ascending branch and anterior branch are two extensions of the lateral sulcus ascending to the frontal lobe.
Surrounding them is the inferior frontal gyrus in whose opercular part lies the motor speech center (Broca).

Lateral sulcus, posterior branch
Telencephalon   2
TA Latin: *Sulcus lat., R. post.*
TA English: *Lateral sulcus, posterior ramus*
Extension from the lateral sulcus stretching into the parietal lobe and which is surrounded by the supramarginal gyrus. The secondary somatosensory cortical fields are located here.

Lateral terminal nucleus
Diencephalon   1
TA Latin: *Nucl. lat. accessorii tracti optici*
TA English: *Lateral nucleus of accessory nuclei of optic tract*
Nuclear region medial to the medial colliculus, on the dorsal margin of the cerebral peduncle.
Is a component of the accessory optic system and involved in coupling of visual information and head movement.

Lateral thalamic nuclei

Diencephalon 2

The lateral nuclear group of the thalamus comprises four nuclear regions:
– posterior lateral nucleus,
– dorsal lateral nucleus,
– anterior thalamic nucleus,
– thalamic pulvinar.

These nuclei are involved in language generation, integrative somatosensory control, pain conduction, gustation and visual processing.

Lateral trigeminothalamic tract

Myelencephalon 3

This fiber bundle conducts information from the spinal nucleus of the trigeminal nerve to the ventral posteromedial thalamic nucleus, which in turn projects to the postcentral gyrus. The fibers emerge from the caudal region, changing immediately to the contralateral side and passing on in the spinothalamic tract to the thalamus. Information about tactile stimulation from the entire facial skin and the lips is carried in this tract.

Lateral ventricle

Meninges & Cisterns 3

TA Latin: *Ventriculus lat.*
TA English: *Lateral ventricle*

Lateral ventricle = 1+2 ventricles of brain.
Situated deep in the hemispheres, composed of the central part and three "horns": anterior horn, posterior horn, inferior horn. A large choroid plexus is situated on the floor of the lateral ventricle. CSF is produced here. CSF flows through the third ventricle to the fourth ventricle, where it flows into the subarachnoid space (cisterns).

Lateral ventricle, anterior horn

Meninges & Cisterns 2

TA Latin: *Ventriculus lat., cornu frontale*
TA English: *Lateral ventricle, frontal horn*

The anterior horn of the lateral ventricle stretches far into the frontal lobe, terminating between the genu of the corpus callosum and the head of the caudate nucleus. More centrally, the two anterior horns are separated by the setpum pellucidum.

Lateral ventricle, central part

Meninges & Cisterns 2

TA Latin: *Ventriculus lat., pars centralis*
TA English: *Lateral ventricle, central part*

The roof of the central segment is formed by the corpus callosum. The choroid plexus of the fourth ventricle stretches along the floor.

In the anterior segment, the left and right central portions are separated by the septum pellucidum and by the splenium of corpus callosum in the posterior segment.

Lateral ventricle, inferior horn

Meninges & Cisterns 2

TA Latin: *Ventriculus lat., cornu temporale*
TA English: *Lateral ventricle, temporal horn*

This section of the lateral ventricle reaches far into the temporal lobe and runs in the immediate vicinity of the hippocampus, which has the finger-shaped evaginations into the inferior horn (digitationes hippocampi)

The CSF-producing choroid plexus of the lateral ventricle (tenia of fornix) reaches into the deepest extensions of this ventricle segment.

Lateral ventricle, posterior horn

Meninges & Cisterns 2

TA Latin: *Ventriculus lat., cornu occipitale*
TA English: *Lateral ventricle, occipital horn*

The posterior horn of the lateral ventricle reaches far into the occipital lobe. On its outer side, the optic radiation runs from the lateral geniculate body to the visual cortex.

On its inner side, the radiation of the corpus callosum is centrally located, with the calcarine sulcus with area 17 (striate cortex) and band of Gennari being encountered on proceeding further into the occipital region.

Lateral vestibular nucleus (Deiters)

Pons 2

TA Latin: *Nucl. vestibularis lat. (Deiters)*
TA English: *Lateral vestibular nucleus (Deiters)*

The lateral vestibular nucleus provides for close coupling of vestibular nuclei with the cerebellum and can be viewed as being an outpost cerebellar nucleus. The large cells have afferents from Purkinje cells of the vermis cerebelli, posterior spinocerebellar tract as well as the auditory tract. Efferents go to the motoneurons of the cervical cord, eye muscle nuclei, red nucleus and the sensory cells in the labyrinth.

Lateral vestibulospinal tract

Medulla spinalis   2

TA Latin: *Tractus vestibulospinalis lat.*
TA English: *Lateral vestibulospinal tract*

The motor fibers from the lateral vestibular nucleus pass in the lateral vestibulospinal tract through the medial longitudinal fasciculus of the spinal cord to the motor anterior horn cells of the sacral cord.

The motor fibers of the medial vestibular nucleus and inferior vestibular nucleus course in the medial vestibulospinal tract.

Lateroinferior thalamic branch

Vessels

TA Latin: *A. choroidea ant.,*
Rr. Capsulae internae
TA English: *Anterior choroidal artery,*
branches to internal capsule

→ Anterior choroid artery, internal capsular branches

Lateromedial occipital artery

Vessels

TA Latin: *A. occipitalis lat./med.*
TA English: *Lateral/medial occipital artery*

→ Posterior cerebral artery, terminal part (cortical)

Lemniscal layer of the superior colliculus

Mesencephalon

→ Superior colliculus, lemniscal layer

Lemniscal trigone   1

TA Latin: *Trigonum lemnisci lat.*
TA English: *Trigone of lateral lemniscus*

The afferent fibers of the lateral lemniscus arrive at the inferior colliculus here, entering this nucleus.

Lemniscus

Pathways   3

TA Latin: *Lemniscus*
TA English: *Lemniscus*

Lemniscus conducts protopathic and epicritic sensibility from the spinal cord and brainstem to the appropriate synaptic centers in the thalamus

(medial lemniscus). The lateral lemniscus is part of the auditory tract.

Lenticular fasciculus

Diencephalon   3

TA Latin: *Fasciculus lenticularis*
TA English: *Lenticular fasciculus*

Fiber tract of the subthalamus. The lenticular fasciculus and ansa lenticularis together form the pallidothalamic projection, the biggest efferent of the globus pallidus. The fibers terminate in the ventral lateral thalamic nucleus, which in turn projects to parts of premotor cortex (area 6) and of the supplementary motor area. They arrive at the thalamic nuclei via the thalamic fasciculus.

Lentiform nucleus

Telencephalon   1

TA Latin: *Nucl. lentiformis*
TA English: *Lentiform nucleus*

Older literature often uses the expression lentiform nucleus for the combination of putamen and globus pallidus. However, since these structures belong together neither ontogenetically nor functionally, use of this expression should be avoided.

Leptomeninges

Meninges & Cisterns   3

TA Latin: *Leptomeninges*
TA English: *Leptomeninges*

Leptos (Greek) = tender, soft; meninx
Leptomeninx is the white cranial meninx. It is composed of the arachnoid resting on the dura mater and the pia mater overlying the brain tissue.

The subarachnoid space opens up between these two layers.

LGB

Diencephalon
TA Latin: *CGL*
TA English: *LGB*

→ Lateral geniculate body (LGB)

Limbic lobe

Telencephalon   3

TA Latin: *Lobus limbicus*
TA English: *Limbic lobe*
Lobes of the limbic system.
Visible only in median section. Formed by the parahippocampal gyrus, cingulate gyrus and hippocampus.

Limbic system

General CNS   3
TA Latin: *Limbisches System*
TA English: *Limbic system*
The limbic system lies deep in the cerebrum and plays an important role in emotions, memory formation, drive and motivation. It comprises the following structures:
– hippocampus,
– fornix,
– amygdaloid body,
– mammillary body,
– parahippocampal gyrus,
– thalamic components.

Limen insula

Telencephalon   2
TA Latin: *Limen insulae*
TA English: *Limen insulae*
Here the lateral surface of the insula joins the basal surface.

Lingula of cerebellum

Cerebellum   2
TA Latin: *Lingula cerebelli*
TA English: *Lingula of cerebellum*
A part of the vermis cerebelli situated on the superior medullary velum, the roof of the fourth ventricle. Belongs to the anterior lobe.
Like the entire vermis cerebelli, the lingula too receives afferents primarily from the spinal cord. It is part of the spinocerebellum = palaeocerebellum.

Lissauer

Medulla spinalis   2
→ **Dorsolateral fasciculus of spinal cord (Lissauer)**

Locus coeruleus

Mesencephalon   3
TA Latin: *Locus caeruleus*
TA English: *Locus caeruleus*

The locus coeruleus belongs to the monoaminergic cell groups, lies in the upper angle of the fourth ventricle and contains approximately half of all noradrenergic cells (A6). It is part of the lateral reticular formation, its efferents reach many parts of the brain and leave the nucleus via the dorsal noradrenergic bundle.
Afferents come primarily from the dorsal raphe nucleus. Due to its pronounced efferents and the facilitating effect of the catecholamines, an important function in regulation of alertness and observational skills is attributed to the locus coeruleus.

Locus coeruleus complex

Mesencephalon   1
The locus coeruleus complex is composed of the actual locus coeruleus

Long central artery (Heubner´s)

Vessels   1
TA Latin: *A. striata med. distalis (Heubneri)*
TA English: *Distal medial striate artery (Heubneri)*
Arises from the postcommunical part of the anterior cerebral artery, shortly after the division of the anterior communicating artery.
It enters the perforated substance and, endowed with several branches, it supplies the caudate nucleus, lentiform nucleus as well as parts of the internal capsule.
Obstruction of the long central artery results in hemiparesis, paralytic symptoms in the tongue and facial musculature as well as in aphasia.

Long gyrus of insula

Telencephalon   2
TA Latin: *Gyrus longus insulae*
TA English: *Long gyrus of insula*
The function of this region is not really known, but viscerosensory and visceromotor functions are suspected.

Longitudinal catecholaminergic bundle

Pathways
TA Latin: *Tractus tegmentalis centralis*
TA English: *Central tegmental tract*
→ **Central tegmental tract**

Longitudinal fissure of cerebrum
Telencephalon   3
TA Latin: *Fissura longitudinalis cerebri*
TA English: *Longitudinal cerebral fissure*
Inter-hemisphere fissure.

It separates the two hemispheres from each other. Deep in the fissure is located the corpus callosum. Encountered here are the large fiber bundles exchanging information across the two hemispheres.

Tumor or hemorrhage in the longitudinal fissure of cerebrum generally triggers symptoms in both body halves. Accordingly, flaccid paralysis of both legs can be induced by a pathological event at the level of area 4. A tumor at the level of area 4 can cause complete blindness.

Longitudinal stria
Telencephalon   2
TA Latin: *Stria longitudinalis*
TA English: *Longitudinal stria*
The longitudinal stria is divided into the:
– lateral longitudinal stria,
– medial longitudinal stria.

The two fiber bundles running on the roof of the corpus callosum contain the supracallosal part of the fornix as well as afferents and efferents of the induseum griseum, which runs between them, and together with it form the hippocampus, supracommissural part, hence are a component of the hippocampus.

Lunate sulcus
Telencephalon   2
TA Latin: *Sulcus lunatus*
TA English: *Lunate sulcus*
Sulcus on the posterior portion of the occipital lobe. Here runs the border between the two Brodmann areas 18 and 19, which jointly form the secondary visual cortex. Recognition and interpretation of complex visual stimuli are effected here.

M

M-region of Holstege et al.
Mesencephalon 1
In the lateral tegmentum of pons there are 2 separate regions for regulating micturition. One region, the M-region of Holstege, is located in the dorsomedial portion of the tegmentum of pons. The other one lies in the ventrolateral tegmentum of pons and has been designated by Holstege as the L-region.

M1-region
Telencephalon
TA Latin: *Pars spenoidalis a. cerebri mediae*
TA English: *Sphenoid part of middle cerebral artery*
→ Precentral gyrus (area 4)

M2-region
Telencephalon
TA Latin: *Rr. terminales a. cerebri mediae*
TA English: *Terminal branches of middle cerebral artery*
→ Motor cortex, supplementary

Magnocellular preoptic nucleus
Diencephalon 1
Near the anterior commissure are situated nuclear groups with large cells whose efferents pass on to the olfactory bulb without synapsing.

Major portion of the trigeminal nerve
TA Latin: *N. trigeminus, radix sensoria*
TA English: *Sensory root of trigeminal nerve*
→ Sensory root of trigeminal nerve (V)

Mammillary body
Diencephalon 3
TA Latin: *Corpus mamillare*
TA English: *Mammillary body*

The mammillary nuclei are located in the medial zone of the hypothalamus. Major afferents arrive via the fornix of hippocampus, while efferents pass largely via the mammillothalamic fasciculus, Vicq d'Azyr bundle to the anterior thalamic nucleus or via the dorsal longitudinal fasciculus (Schütz) to the visceral centers in the brainstem and spinal cord.
Component of the Papez neuronal circuit. Involved in affective actions and learned processes.
Damage to the mammillary body, e.g. in the case of alcoholic encephalopathy, results in affective impairments and marked loss of perceptivity.

Mammillary body, lateral nucleus
Diencephalon 1
TA Latin: *Nucl. mammillaris lat.*
TA English: *Lateral nucleus of mammillary body*
→ Mammillary nuclei

Mammillary body, medial nucleus
Diencephalon 1
TA Latin: *Nucl. mammillaris med.*
TA English: *Medial nucleus of mammillary body*
→ Mammillary nuclei

Mammillary nuclei
Diencephalon 2
TA Latin: *Nuclei mamillares*
TA English: *Nuclei of mammillary body*
A distinction is made between the following nuclei of the mammillary body:
– mammillary body, medial nucleus,
– intermediate mammillary nucleus,
– mammillary body, lateral nucleus,
– posterior nucleus.
The medial nuclear region is especially pronounced in humans and connected via the thalamus with the prefrontal cortex.
Selective dysfunction of the mammillary body, medial nucleus results in Korsakoff syndrome (amnetic syndrome with anterograde and retrograde impaired memory, and diminished drive).

Mammillotegmental fasciculus
Pathways
TA Latin: *Fasciculus mammillotegmentalis*

TA English: *Mammillotegmental fasciculus*
→ Mammillotegmental tract

Mammillotegmental tract
Diencephalon   2
TA Latin: *Fasciculus mammillotegmentalis*
TA English: *Mammillotegmental fasciculus*
Emerges together with the mammillothalamic
fasciculus, Vicq d'Azyr bundle from the principle mammillary fasciculus and passes to the
tegmentum of mesencephalon, where it terminates predominantly in two nuclei:
– dorsal tegmental nucleus (Gudden),
– tegmental pontine reticular nucleus
(Bechterew)

Mammillothalamic fasciculus, Vicq d'Azyr bundle
Pathways
TA Latin: *Fasciculus mammillothalamicus*
TA English: *Mammillothalamic fasciculus, Vicq d'Azyr bundle*
→ Mammillothalamic tract

Mammillothalamic tract
Diencephalon   3
TA Latin: *Fasciculus mammillothalamicus*
TA English: *Mammillothalamic fasciculus*
The principle mammillary fasciculus conducts
the efferents of the mammillary body somewhat
dorsally, then dividing into the mammillothalamic fasciculus, Vicq d'Azyr bundle and the
mammillotegmental tract. The former passes to
the anterior thalamic nucleus. It forms part of
the Papez neuronal circuit, which plays an important part in emotion and memory formation

Mandibular branch of the trigeminal nerve
Nerves
TA Latin: *R. mandibularis n. trigemini*
TA English: *Mandibular branch of trigeminal nerve*
→ Mandibular nerve (V3)

Mandibular nerve (V3)
Nerves   3
TA Latin: *N. mandibularis (N.V3)*
TA English: *Mandibular nerve (V3)*

The greatest branch of the trigeminal nerve divides into a number of sensory (4) and motor
(>4) nerves. It provides for sensory innervation
of the skin of the lower jaw, up to the posterior
temples, the mandible with all teeth, the anterior
2/3 of the tongue and the inside of the cheeks. It
provides motor innervation for the entire
masticatory muscles (masseter, temporal, floor
of the mouth muscles).
Skull: oval foramen of sphenoid bone.

Marginal cells
Medulla spinalis   2
TA Latin: *Cornu posterius, Nucleus marginalis*
TA English: *Posterior horn, marginal nucleus*
Dorsal margin of the posterior horn. The cells
project to the contralateral mesencephalic
reticular formation and the contralateral thalamus. Afferents come from the proprioceptive
and nociceptive receptors.

Marginal sinus
Vessels   3
TA Latin: *Sinus marginalis*
TA English: *Marginal sinus*
The two marginal sinuses arise from the division
of the occipital sinus and transport its blood fur-
ther in the direction of the bilateral superior
bulb of the jugular vein.

Margo liber falcis cerebri
Meninges & Cisterns   2
The free end of the falx cerebri. The inferior
sagittal sinus courses in this end.

Mastoid foramen
Meninges & Cisterns   1
TA Latin: *Foramen mastoideum*
TA English: *Mastoid foramen*
Point of passage of the occipital artery, mastoid
branch, situated on the mastoid process.

Maxillary artery
Vessels   3
TA Latin: *A. maxillaris*
TA English: *Maxillary artery*
Like the facial artery, it arises from the external
carotid artery. It supplies the tonsils,
masticatory muscles, nasal cavity, gums, mandi-
ble, teeth, maxilla, temporomandibular joint,

auditory meatus, middle ear, tympanic cavity and meninges. It anastomoses with the facial artery via the infraorbital artery.

Maxillary artery, mandibular part
Vessels   2
The maxillary artery emerges from the external carotid artery. The pterygoid part joins the mandibular part.

Maxillary artery, pterygoid part
Vessels   2
→ Maxillary artery, mandibular part

Maxillary branch of the trigeminal nerve
Nerves
TA Latin: *R. maxillaris n. trigemini*
TA English: *Maxillary branch of trigeminal nerve*
→ Maxillary nerve (V2)

Maxillary nerve (V2)
Nerves   3
TA Latin: *N. maxillaris (N.V2)*
TA English: *Maxillary nerve (V2)*
Like ophthalmic nerve (V1), the maxillary nerve is also purely sensory. In the pterygopalatine fossa it divides into the ganglion branches, the zygomatic nerve and the infraorbital nerve. Sensory innervation is provided via these nerves..
– for parts of the nasal mucosa,
– for parts of the palatine mucosa,
– for the upper jaw together with the upper teeth,
– for the facial skin beneath the eye up to the anterior temple.
Skull: Foramen rotundum.

Maxillary veins
Vessels   2
TA Latin: *Vv. maxillares*
TA English: *Maxillary veins*
The maxillary veins carry blood from the pterygoid sinus into the retromandibular vein. Their catchment area includes the tonsils, soft palate, masticatory musculature, nasal and sinus cavities, teeth, jaws, calvaria and dura mater.

Medial accessory nucleus of the olive
Myelencephalon   2
TA Latin: *Nucl. olivaris accessorius med.*
TA English: *Medial accessory olivary nucleus*
The two accessory olives (dorsal accessory nucleus of the inferior olive and medial accessory nucleus of the inferior olive) receive afferents from the central gray matter of mesencephalon and the spinal cord (via spino-olivary tract) and project to the cerebellum, to the following parts: interpositus nucleus (emboliform +globose nuclei), fastigial nucleus, vermis cerebelli and cerebellar hemisphere, intermediate part.
The accessory olives are also involved in movement coordination.

Medial amygdaloid nucleus
Telencephalon   1
TA Latin: *Nucl. amygdalae med.*
TA English: *Medial amygdaloid nucleus*
→ Amygdaloid body

Medial and lateral preoptic nuclei
Diencephalon   1
TA Latin: *Nuclei preoptici med. + lat.*
TA English: *Medial and lateral preoptic nuclei*
The lateral preoptic nucleus is involved in locomotion.
The medial preoptic nucleus, conversely, is involved in thermoregulation, hypovolemic thirst, male sexual behavior, nursing care, modulation of gonadotropin secretion.

Medial arcuate ligament   1
TA Latin: *Ligamentum arcuatum med.*
TA English: *Medial arcuate ligament*
Arcuate ligament of the diaphragm, penetrated by the azygos vein or the hemiazygos vein.

Medial atrial vein
Vessels   1
A vein descending to the great cerebral vein and coursing along the inner wall of the lateral ventricle.

Medial bundle of the forebrain
Pathways
TA Latin: *Fasciculus med. telencephali*
TA English: *Medial forebrain bundle*

→ Medial telencephalic fasciculus,
(medial bundle of forebrain)

Medial central nucleus

Diencephalon

TA Latin: *Nucl. centralis med.*

TA English: *Central medial nucleus*

The intralaminar nuclei of the thalamus can be devided in two groups: the caudal group (centromedian nucleus and parafascicular nucleus) and the rostral group (central lateral nucleus, paracentral nucleus and central medial nucleus).

The nuclei play an import role in the perception of pain and they project to cortex and striatum.

Medial cuneate nucleus

Myelencephalon   2

Part of the dorsal column.

The medial cuneate nucleus is strictly speaking the cuneate nuclei. Epicritic afferents synapse here and project to the thalamus.

Conversely, in the lateral cuneate nucleus, afferents enter from the vestibular system (equilibrium). This region projects to the cerebellum.

Medial eminence

Pons   1

TA Latin: *Eminentia medialis*

TA English: *Medial eminence*

Protrusion created by underlying cranial nerve nuclei and pathways.

Medial frontal gyrus

Telencephalon   3

TA Latin: *Gyrus front. medius*

TA English: *Middle frontal gyrus*

The medial frontal gyrus has no clearly identified task like the other two frontal gyri. Together with the frontal portions of the superior frontal gyrus and inferior frontal gyrus it is probably responsible for higher psychic and mental tasks. It is categorized like the other frontal gyri as belonging to the association area.

Bilateral damage to the frontal association areas due to large-surface tumors, hemorrhage or degeneration leads to marked diminishment of intellectual faculties. Drive, concentration power and social control mechanisms are greatly reduced, concomitantly resulting in slowing down of all activities.

Medial frontobasal artery

Vessels   2

TA Latin: *A. frontobasalis med.*

TA English: *Medial frontobasal artery*

Arises from the postcommunical part of the anterior cerebral artery at the level of the genu of the corpus callosum and passes at the base of the brain in the direction of the frontal lobe, where it can also be seen superficially.

Medial geniculate body, dorsal part

Diencephalon   1

The MGB consists of three regions:

– medial geniculate body, dorsal part
– medial geniculate body, medial part
– medial geniculate body, ventral part

The dorsal part of the MGB receives fibers from the pericentral nucleus of the inferior colliculus and is connected with the auditory association cortex in the temporal plane and superior temporal gyrus.

Medial geniculate body, medial part

Diencephalon   1

The MGB consists of three regions:

– medial geniculate body, dorsal part
– medial geniculate body, medial part
– medial geniculate body, ventral part

The medial magnocellular portion receives collaterals from the medial lemniscus and spinothalamic tract. There are also diffuse connections to the auditory region and contiguous regions.

Medial geniculate body (MGB)

Diencephalon   3

TA Latin: *Corpus geniculatum med. (CGM)*

TA English: *Medial geniculate body (MGB)*

The MGB is an important synaptic center of the auditory tract and is subdivided into three regions:

– medial geniculate body, dorsal part
– medial geniculate body, medial part
– medial geniculate body, ventral part

The morphologic correlate of the tonotopic organization is the laminar structure. Efferents pass to the planum temporale and superior temporal gyrus, afferents originate primarily in the inferior colliculus.

Medial geniculate nucleus, ventral part

Diencephalon   1

The MGB consists of three portions:
– medial geniculate body, dorsal part
– medial geniculate body, medial part
– medial geniculate body, ventral part

The ventral portion features tonotopic organization, receiving afferents from the central nucleus of the inferior colliculus and projecting (likewise tonotopically) to the temporal operculum.

Medial habenular nucleus

Diencephalon   2

TA Latin: *Nucl. habenularis med.*
TA English: *Medial habenular nucleus*

The medial habenular nucleus belongs to the epithalamus and is composed of small, densely packed, oval cells. Afferents come via the medullary stria from the septal diagonal band , from mesencephalic raphe nuclei and from the ventral portion of the central gray matter of mesencephalon. Efferents to the interpeduncular nucleus, to the dorsal raphe nuclei and the superior central raphe nuclei. The exact functions have not been clarified.

Medial inferior artery of cerebellar hemisphere, medial branch

Vessels

TA Latin: *A. inf. ant. cerebelli*
TA English: *Anterior inferior cerebellar artery*

→ Anterior inferior cerebellar artery

Medial lemniscus

Pathways   3

TA Latin: *Lemniscus med.*
TA English: *Medial lemniscus*

The medial lemniscus arises from the union of the spinothalamic tract (anterior column of the spinal cord, protopathic sensibility), the bulbothalamic tract (gracile nucleus and cuneate nucleus, epicritic sensibility), the trigeminal lemniscus (face) and afferents of the solitary nucleus (gustatory sensibility). Hence it is also called the somatosensory tract. The fibers terminate in the corresponding thalamic nuclei,

which in turn project to the somatosensory cortex.

Medial longitudinal fasciculus

Pathways   3

TA Latin: *Fasciculus longitudinalis med.*
TA English: *Medial longitudinal fasciculus*

The medial longitudinal fasciculus is composed of two fiber components:
–   vestibular component: conducts efferents of the vestibular nuclei to the nuclei for controlling eye and cervical muscles, thus coordinating organ of equilibrium with eye and head movements.
–   internuclear component: it coordinates motor cranial nerve nuclei and provides for synchronous eye movement and coordination of pharyngeal muscles while speaking and swallowing.

The following tracts course here:
– interstitiospinal tract,
– reticulospinal tract,
– tectospinal tract,
– lateral vestibulospinal tract,
– medial vestibulospinal tract.

Medial mammillary nucleus   2

TA Latin: *Nucl. mammillaris med.*
TA English: *Medial nucleus of mammillary body*

→ Mammillary body, medial nucleus

Medial medullary lamina

Telencephalon   2

TA Latin: *Lamina medullaris med.*
TA English: *Medial medullary lamina*

Layer of white matter separating the globus pallidus into a globus pallidus, lateral part and a globus pallidus, medial part.

Medial nucleus of the superior olive

Mesencephalon   2

TA Latin: *Nucl. olivaris sup. med.*
TA English: *Medial superior olivary nucleus*

Is a relay nucleus of the auditory tract. Afferents come from the ventral cochlear nucleus. Efferents go to the ipsi- and contralateral inferior colliculus, central nucleus.

Medial occipitotemporal gyrus
Telencephalon   3
TA Latin: *Gyrus occipitotemporalis med.*
TA English: *Medial occipitotemporal gyrus*
→ Lateral occipitotemporal gyrus

Medial olfactory stria
Telencephalon   3
TA Latin: *Stria olfactoria med.*
TA English: *Medial olfactory stria*
At the brain end, the olfactory bulb divides into the lateral stria and the medial stria.
The medial stria passes in the direction of the subcallosal area where, inter alia, the anterior olfactory nucleus, is located.

Medial parabrachial nucleus
Diencephalon   2
TA Latin: *Nucl. parabrachialis med.*
TA English: *Medial parabrachial nucleus*
This nucleus plays a role in forwarding gustatory information to the thalamus and in further processing of impulses from the respiratory tract as well as cardiovascular and gastrointestinal areas, ascending from the caudal segment of the solitary nucleus. Efferents go to the ventral posteromedial nucleus, intralaminar nuclei, lateral hypothalamic area, central amygdaloid nucleus, basal nucleus and to the cerebral cortex.

Medial posterior choroid artery
Vessels
TA Latin: *A. cerebri post., R. choroideus post. med.*
TA English: *Posterior cerebral artery, posterior medial choroidal branch*
→ Posterior cerebral artery, medial posterior choroid branch

Medial preoptic nucleus (POML)
Diencephalon   2
TA Latin: *Nucl. preopticus med.*
TA English: *Medial preoptic nucleus*
Lies in the preoptic area, involved in thermoregulation, hypovolemic thirst, male sexual behavior, brood care, modulation of gonadotropin secretion. Larger afferents from the amygdala, subiculum, interstitial nucleus of the stria terminalis, lateral septal nucleus, insula. Efferents to the diagonal band, septum, substantia innominata, interstitial nucleus, amygdaloid body, brainstem. Has numerous cells with receptors for gonadal steroids.
Disruption of this nucleus induces ongoing hypothermia, while stimulation results in hyperthermia.
The menstruation cycle is also interrupted in the event of dysfunctioning of this area.

Medial pretectal nucleus
Mesencephalon   1
→ Pretectal area

Medial reticular formation
Pons   2
The magnocellular, medial region of the RF stretches across the entire brainstem and enables the following sections to be distinguished:
– gigantocellular reticular nucleus,
– caudal pontine reticular nucleus,
– oral pontine reticular nucleus.
Belonging here are also the mesencephalic nuclei: cuneiform nucleus, subcuneiform nucleus and central medulla oblongata nucleus. This region of the RF is involved in motor and sensory tasks.

Medial septal nucleus (Ch1)
Telencephalon   2
TA Latin: *Nucl. septalis med. (Ch1)*
TA English: *Medial septal nucleus (Ch1)*
Belongs to the septal nuclei.
Receives its biggest afferents from the lateral septal nucleus, preoptic area, mammillary body, medial nucleus and dorsal tegmental nucleus.
Large efferents to hippocampus, including subiculum, and enterorhinal cortex.
Smaller efferents to the preoptic area, mammillary complex, tegmental area and raphe nuclei.

Medial superior olivary nucleus
TA Latin: *Nucl. olivaris sup. med.*
TA English: *Medial superior olivary nucleus*
→ Medial nucleus of the superior olive

Medial tegmental tract
Pathways   1

This tract conducts impulses from the central gray matter of mesencephalon to the nucleus of the superior olive and is likewise involved in movement coordination.

Medial telencephalic fasciculus
Pathways 2
TA Latin: *Fasciculus med. telencephali*
TA English: *Medial forebrain bundle*
"Tract of the lateral paracore".
The medial telencephalic fasciculus or the medial forebrain bundle can be viewed as the central longitudinal pathway of the limbic forebrain-midbrain axis. This bundle is the union of loosely arranged thin fibers, which stretch from the septal area to the tegmentum of mesencephalon.

Medial terminal nucleus
Diencephalon 1
TA Latin: *Nucl. med. accessorii tracti optici*
TA English: *Medial nucleus of accessory nuclei of optic tract*
Nuclear region near the medial margin of the substantia nigra.
Is a component of the accessory optic system and involved in coupling of visual information and head movement.

Medial thalamic nucleus
Diencephalon 2
TA Latin: *Nucl. med. thalami*
TA English: *Medial thalamic nucleus*
The medial thalamic nucleus has two parts:
1) the medial magnocellular part (medial thalamic nucleus, medial part) has reciprocal connections with the olfactory cortical areas. It also has fibers from the amygdaloid body and temporal pole.
2) the lateral parvocellular part is connected with the frontal eye field and the prefrontal cortex. Afferents from the superior colliculus and substantia nigra, vestibular complex and tegmentum area.

Medial vestibular nucleus
Pons 2
TA Latin: *Nucl. vestibularis med.*
TA English: *Medial vestibular nucleus*
→ Vestibular nuclei (medial, superior, inferior)

Medial vestibulospinal tract
Medulla spinalis 2
TA Latin: *Tractus vestibulospinalis med.*
TA English: *Medial vestibulospinal tract*
The efferents going from the medial vestibular nucleus and inferior vestibular nucleus in the direction of the spinal cord form the medial vestibulospinal tract which passes in the medial longitudinal fasciculus of the spinal cord into the cervical and upper thoracic cord, ending here on the motoneurons, innervating the cervical musculature and the upper extremities.
The motor fibers of the lateral vestibular nucleus runs in the lateral vestibulospinal tract.

Median and lateral apertures of fourth ventricle
Meninges & Cisterns 3
TA Latin: *Aperturae (mediana + lat.) ventriculi quarti*
TA English: *Median and lateral apertures of fourth ventricle*
The aperture of the fourth ventricle is a point where cerebrospinal fluid (CSF) passes from the ventricular system, the fourth ventricle, into the subarachnoid system, the cerebellomedullary cistern.
Exchange of CSF between the subarachnoid system and the blood circulation is effected in the arachnoid granulations and in the evaginations of the spinal dura mater at the spinal nerves.

Median artery of corpus callosum
Vessels 2
TA Latin: *A. callosa mediana*
TA English: *Median callosal artery*
An artery that branches at the level of the anterior communicating artery, passing by the anterior commissure to the genu of the corpus callosum and supplying the latter.

Median eminence
Diencephalon 1
TA Latin: *Eminentia mediana*
TA English: *Median eminence*
An elevation that is only poorly developed in humans at the attachment of the infundibulum, caused by vessels growing into the wall of the infundibulum.

Median preoptic nucleus (POMN)
Diencephalon   1
TA Latin: *Nucl. preopticus medianus*
TA English: *Median preoptic nucleus*
This nucleus which is also abbreviated to medial
preoptic nucleus is closely coupled to the
organum vasculosum of the lamina terminalis
and the subfornical organ. Efferents pass on to
the supraoptic nucleus.

Median thalamic nucleus
Diencephalon   2
TA Latin: *Nuclei mediani thalami*
TA English: *Median nuclei of thalamus*
A small cell cluster situated at the midline, also
called "nuclei of the midline". Not to be con-
fused with the medial thalamic nucleus.
It includes:
– rhomboid nucleus,
– reuniens nucleus (projects to hippocampus;
afferents from parabrachial nuclei; integrated in
the information chain between the visceral periphery to the limbic system).
– paratenial nucleus (projections to the striatal
body).

Medulla oblongata
Myelencephalon   3
TA Latin: *Medulla oblongata*
TA English: *Medulla oblongata*
→ Myelencephalon

Medulla spinalis
Medulla spinalis   3
TA Latin: *Medulla spinalis*
TA English: *Medulla spinalis*
→ Spinal cord

Medullary body of cerebellum
Cerebellum   2
TA Latin: *Corpus medullare cerebelli*
TA English: *White substance of cerebellum*
The large white matter situated in the cerebellar
hemispheres is called the medullary body of cerebellum. It is composed largely of afferent and
efferent fibers and accommodates the central
cerebellar nuclei, above all the large dentate nucleus with its saw–toothed shape. Due to the at-
tractive arboreal structure depicted in the me-

dian section, the cerebellar white matter is called
arbor vitae (the tree of life).

Medullary branches
Vessels   1
TA Latin: *Rr. medullares*
TA English: *Medullary branches*
At the level of the pons, i.e. shortly before the
junction of the vertebral arteries and basilar artery, two small branches are often given off, entering the myelencephalon via the foramen ce-
cum, and hence called medullary branches. Here
one distinguishes between the anteromedial
medullary branches and the posteromedial
medullary branches.

Medullary cistern
Meninges & Cisterns   2
Cisterns are also formed in the spinal canal in
the subarachnoid space, with these being globally known as the medullary cistern. They join
the cerebromedullary cistern at the foramen
magnum.

Medullary stria of the fourth ventricle
Pons   1
TA Latin: *Stria medullaris ventriculi quarti*
TA English: *Medullary stria of fourth ventricle*
Myelinated fibers which on the floor of the
fourth ventricle (rhomboid fossa) pass outside
across the medial eminence and the vestibular
area and enter the inferior cerebellar peduncle.
These fibers may have their source in the arcuate
nucleus.
The fibers mark the borderline between pons
and myelencephalon, but have heterogenous expression.

Medullary stria of the thalamus
Diencephalon   2
TA Latin: *Stria medullaris thalami*
TA English: *Stria medullaris of thalamus*
The medullary stria of the thalamus are fiber
bundles on the inside of the thalamus complex
in which course fibers from the hypothalamus,
preoptic area and the septum to the habenula.
The fiber bundle is a component of the limbic
system.

Meningeal branch of the 10th thoracic nerve
Meninges & Cisterns 1
On exiting via the intervertebral foramen, the thoracic nerve divides into several branches, including also the meningeal branch that provides sensory innervation for the spinal meninges.

Meningeal branch of the posterior vertebral artery
Vessels
TA Latin: *R. meningeus post. Arteriae vertebralis*
TA English: *Meningeal branch of posterior vertebral artery*
→ Vertebral artery, intracranial part, posterior meningeal branch

Meningeal vein
Vessels 2
TA Latin: *V. meningea*
TA English: *Meningeal vein*
Veins from the cranial meninges.

Mesencephalic aqueduct
Meninges & Cisterns 3
TA Latin: *Aquaeductus mesencephali*
TA English: *Aqueduct of midbrain*
The mesencephalic aqueduct is a duct between the third ventricle and the fourth ventricle.
The duct leads transversely through the mesencephalon, past the quadrigeminal lamina and mammillary body.

Mesencephalic arteries
Vessels
TA Latin: *Aa. mesencephalicae*
TA English: *Mesencephalic arteries*
→ Quadrigeminal artery

Mesencephalic "attack areas"
Mesencephalon 1
Small nuclear region in the mesencephalon whose stimulation results in the formation of somatomotor attack potential, without triggering the customary concomitant autonomic mechanisms.
Hence one can distinguish between autonomic and somatomotor "attack areas".

Mesencephalic locomotor area
Mesencephalon 1
Mesencephalic nuclear region that can coordinate simple step movements. Corresponds to the pedunculopontine tegmental nucleus, pars compacta.

Mesencephalic nucleus of the trigeminal nerve
Mesencephalon 2
TA Latin: *Nucl. mesencephalicus n. trigemini*
TA English: *Mesencephalic nucleus of trigeminal nerve*
This long nucleus extends through the central gray matter of mesencephalon. It is formed by the perikarya of proprioceptive muscle spindle afferents, which pass via the motor root of the trigeminal nerve (proprioceptive fibers) into the mesencephalon and are the sole primary afferents of the NS having their cell body in the CNS. Efferents are project to the motor nucleus of the trigeminal nerve and the reticular formation.

Mesencephalic reticular formation
Mesencephalon 2
TA Latin: *Formatio reticularis tegmentum mesencephali*
TA English: *Mesencephalic reticular formation*
The gigantocellular portion of the RF, the medial reticular formation, stretches as far as the mesencephalon, forming there a series of nuclei.
– cuneiform nucleus,
– subcuneiform nucleus and
– central nucleus of the myelencephalon.

Mesencephalic tract of the trigeminal nerve
Mesencephalon 2
TA Latin: *Tractus mesencephalicus n. trigemini*
TA English: *Mesencephalic tract of trigeminal nerve*
Fiber bundle containing, on the one hand, the afferent, proprioceptive axons of the mesencephalic nucleus of the trigeminal nerve and, on the other hand, two collaterals to the motor nucleus of the trigeminal nerve. This tract also carries efferents of the nucleus to the motor nucleus of the trigeminal nerve and to the reticular formation.

Mesencephalic veins
Vessels   1
Inconstant venous plexus around the mesencephalon.

Mesencephalon
Mesencephalon   3
TA Latin: *Mesencephalon*
TA English: *Mesencephalon*
Primordial brain tissue around the mesencephalic aqueduct. 3 parts:
1) The cerebral peduncles contain large fiber bundles from the cerebrum.
2) The tegmentum area contains substantia nigra, red nucleus, cranial nerve nuclei and parts of the reticular formation.
3) The tectum is formed from the quadrigeminal plate (inferior and superior colliculi).

Tasks: eye movement, fine motor control, sensory-motor coupling, effector movement, synaptic center.

Mesencephalospinal fibers
Pathways   1
Projections of the central gray matter of mesencephalon to neurons in the spinal cord.

Metathalamus
Diencephalon   3
TA Latin: *Metathalamus*
TA English: *Metathalamus*
Part of the diencephalon. Comprises the lateral geniculate body as well as the medial geniculate body.

Metencephalic reticular formation
Pons   1
TA Latin: *Formatio reticularis metencephali*
TA English: *Metencephalic reticular formation*
Metencephalic reticular formation, at the level of the pons.

Metencephalic tegmental decussation
Myelencephalon   2
TA Latin: *Decussatio tegmentalis metencephali*
TA English: *Tegmental decussation of midbrain*

Here the medial longitudinal fasciculus decussates to the contralateral side.

Metencephalon
General CNS   3
TA Latin: *Metencephalon*
TA English: *Metencephalon*
Pons and cerebellum.

MGB
Diencephalon
TA Latin: *CGM*
TA English: *MGB*
→ Medial geniculate body (MGB)

Middle arterial vein of lateral ventricle
Vessels
→ Medial atrial vein

Middle cerebellar peduncle
Cerebellum   3
TA Latin: *Pedunculus cerebellaris med.*
TA English: *Middle cerebellar peduncle*
Composed exclusively of afferent fibers which all come from the pontine nuclei, and account for the majority of cerebellar afferents and are designated collectively as the pontocerebellar tract. The fibers decussate before entering the peduncle of the contralateral side, giving off collaterals to the dentate nucleus and projecting to the cerebral cortex of the cerebellar hemispheres.

Middle cerebral artery
Vessels   3
TA Latin: *A. cerebri media*
TA English: *Middle cerebral artery*
Emerges together with the anterior cerebral artery from the internal carotid artery and, with a diameter of 4 mm, is the greater branch of the two. Together with the anterior cerebral artery, the two branches constitute the principal arteries of one half of the cerebrum. A distinction is made between the sphenoid part (M1) and insular part (M2).
The supply area comprises the deep and superficial regions throughout the entire telencephalon.
Dysfunction of the middle cerebral artery results

in severe damage generally affecting the entire contralateral body half.

Motor, motosensory and sensory dysfunctioning of the speech center are common.

Middle cerebral artery, frontal trunk
Vessels   3
The frontal trunk (=anterior trunk) is the term used to designate the sum of all frontal branches of the middle cerebral artery. These include important vessels such as the lateral frontobasal artery, artery of precentral sulcus, central sulcus artery, artery of postcentral sulcus, supramarginal artery and artery of the angular gyrus.

Supplied are the frontal lobes, pre- and postcentral gyri, anterior parietal gyri, insula and optic radiation.

Middle cerebral artery, insular part
Vessels   3
TA Latin: *A. cerebri media, pars insularis*
TA English: *Middle cerebral artery, insular part*
Whereas the sphenoid part (M1) supplies primarily the deep regions of the cortex, the insular part (M2), which is situated deep in the lateral sulcus, gives off a variable number of branches to the surface of the cerebral cortex.

A distinction is made between the frontal trunk with the frontal branches, the parietal trunk with the parietal branches as well as the posterior trunk with the occipitotemporal branches.

Middle cerebral artery, parietal trunk
Vessels   1
The parietal trunk is the term used to designate the sum of all parietally oriented branches of the middle cerebral artery.

Middle cerebral artery, posterior trunk
Vessels   3
The posterior trunk (=inferior trunk) is the term used to designate the sum of all occipitotemporal branches of the middle cerebral artery. These include important vessels such as the anterior temporal artery, intermediate temporal artery, posterior temporal artery as well as the insular branches.

Supplied are the temporal lobes, Wernicke's area, cerebral insula, extreme capsule, claustrum, internal capsule, putamen and amygdaloid body.

Middle cerebral artery, sphenoid part
Vessels   3
TA Latin: *A. cerebri media, pars sphenoidalis*
TA English: *Middle cerebral artery, sphenoid part*
Whereas the insular part (M2) supplies large sections of the surface of the cerebral cortex, the sphenoid branches (M1) transport oxygenated blood primarily to the deep regions of the cortex. The anterolateral central arteries supply parts of the rostral commissure, of putamen, of internal capsule and the corona radiata, body of caudate nucleus and part of the head of caudate nucleus.

Middle cerebral artery, striate branches
Vessels   1
TA Latin: *A. cerebri media, Rr. striati*
TA English: *Middle cerebral artery, striate branches*
These branches of the Middle cerebral artery form characteristic cavities in the nervous tissue (Anterior perforated substance).

Middle cerebral artery, temporal trunk
Vessels   1
The middle cerebral artery, posterior trunk or temporal trunk is the term used to designate the sum of all temporally oriented branches of the middle cerebral artery.

Middle cerebral artery, terminal part
Vessels   2
TA Latin: *A. cerebri media, Rr. terminales*
TA English: *Middle cerebral artery, terminal branches*
The terminal part (cortical part) is the term used to designate the sum of all lateral and terminal branches of the anterior and posterior trunks, which together form the insular part. They supply the cerebral insula, opercula, as well as large regions of the cerebral cortex around the insula.

Middle epidural vein
Vessels
Together, the lateral epidural vein and the middle epidural vein form the anterior internal ver-

tebral venous plexus. They drain the spinal cord and spinal meninges.

Middle frontal artery
Vessels
TA Latin: *A. supratrochlearis*
TA English: *Supratrochlear artery*
→ Supratrochlear artery

Middle gray layer of the superior colliculus
Mesencephalon
TA Latin: *Stratum griseum intermedium colliculi sup.*
TA English: *Intermediate grey layer of superior colliculus*
→ Superior colliculus, middle gray layer

Middle intermediate temporal branch
Vessels   1
TA Latin: *A. occipitalis lat., R. Temporalis intermedius*
TA English: *Intermediate temporal branch of lateral occipital artery*
→ Lateral occipital artery, middle intermediate temporal branches

Middle meningeal artery
Vessels   3
TA Latin: *A. meningea media*
TA English: *Middle meningeal artery*
The largest artery of the dura. It arises from the maxillary artery and passes through the spinous foramen into the middle cranial fossa, where it divides into the frontal and parietal branches. Both supply bones, dura and soft tissues of the middle cranial fossa.
Via the anastomotic branch, the frontal branch forms an anastomosis with the lacrimal artery.
Is often affected in the event of trauma to the skull, resulting in hematomas due to epidural hemorrhages.

Middle meningeal artery, anastomotic branch
Vessels   2
TA Latin: *A. meningea media, R. anastomoticus cum a. lacrimali*
TA English: *Middle meningeal artery, anastomotic branch with lacrimal artery*

The part of the middle meningeal artery forming an anastomosis with the lacrimal artery.

Middle meningeal artery, frontal branch
Vessels   2
TA Latin: *A. meningea media, R. front.*
TA English: *Middle meningeal artery, frontal branch*
Arises together with the parietal branch from the middle meningeal artery. The branch divides further and supplies the frontal, temporal and parietal cranial regions (bones, dura, soft tissues).

Middle meningeal artery, frontal branch, intraosseal part
Vessels   1
Inconstant lateral branch of the middle meningeal artery, frontal branch.

Middle meningeal artery, parietal branch
Vessels   2
TA Latin: *A. meningea media, R. parietalis*
TA English: *Middle meningeal artery, parietal branch*
Arises together with the frontal branch from the middle meningeal artery. The branch divides further and supplies the occipital, temporal and parietal cranial regions (bones, dura, soft tissues).

Middle meningeal veins
Vessels   2
TA Latin: *Vv. meningeae mediae*
TA English: *Middle meningeal veins*
Drain venous blood from the dura mater of brain into the pterygoid sinus.

Middle occipital artery
Vessels   3
TA Latin: *A. occipitalis med.*
TA English: *Medial occipital artery*
Together with the lateral occipital artery, it forms the terminal part of the posterior cerebral artery. It passes to the occipital lobe where it divides into the calcarine and parieto-occipital branch. The calcarine branch supplies the medial occipitotemporal gyrus, lower cuneus and

visual cortex. The parieto-occipital branch supplies the upper cuneus, posterior part of the precuneus, as well as parts of the superior parietal lobule.

Middle occipital artery, calcarine branch
Vessels 2
TA Latin: *A. occipitalis med., R. calcarinus*
TA English: *Medial occipital artery, calcarine branch*
The middle occipital artery emerges from the terminal part (P4) of the posterior cerebral artery and enters two main branches: calcarine branch and parieto-occipital branch. The calcarine branch passes deep in the calcarine sulcus to the occipital pole, where it supplies the medial occipitotemporal gyrus as well as the caudal section of the cuneus.

Middle occipital artery, parietal branch
Vessels 1
TA Latin: *A. occipitalis med., R. parietalis*
TA English: *Medial occipital artery, parietal branch*
The middle occipital artery emerges from the terminal part (P4) of the posterior cerebral artery and enters two main branches: calcarine branch and parieto-occipital branch.
On its way to the brain surface, the parieto-occipital branch gives off the parietal branch which supplies middle sections of the parietal lobe.

Middle occipital artery, parieto-occipital branch
Vessels 1
TA Latin: *A. occipitalis med., R. parietooccipitalis*
TA English: *Medial occipital artery, parieto-occipital branch*
The middle occipital artery emerges from the terminal part (P4) of the posterior cerebral artery and enters two main branches: calcarine branch and parieto-occipital branch.
The parieto-occipital branch passes in the parieto-occipital sulcus to the surface of the brain, where it supplies the upper section of the cuneus, parts of the precuneus and superior parietal lobule.

Middle pontine arteries
Vessels
TA Latin: *Aa. pontis, R. med.*
TA English: *Pontine arteries, medial branch*
Small lateral arteries often are given off from the basilar artery into the surrounding tissue of the pons.
Some of these enter the pons more laterally, and some more medially.
Hence a distinction is made between lateral pontine arteries and medial pontine arteries.
Depending on their course, a further distinction can be made between inferolateral and superolateral branches.

Middle temporal gyrus
Telencephalon 3
TA Latin: *Gyrus temporalis medius*
TA English: *Middle temporal gyrus*
The middle temporal gyrus passes in the center over the temporal lobe from the temporal pole to the occipital incision.

Middle temporal veins
Vessels 1
TA Latin: *Vv. temporales med.*
TA English: *Middle temporal veins*
Drain venous blood from the temporal lobe and into the basal vein.

Minor portion of the trigeminal nerve
TA Latin: *N. trigeminus, radix motoria*
TA English: *Motor root of trigeminal nerve*
→ Motor root of trigeminal nerve (V)

Monoaminergic cell groups
General CNS 3
TA Latin: *Cellulae monoaminergicae*
TA English: *Monoaminergic cells*
Monoaminergic neurotransmitters of the CNS: dopamine, noradrenaline, adrenaline and serotonin. Accordingly a distinction is made between:

– dopaminergic cell groups A8-A10,
– noradrenergic cell groups A1-A7,
– adrenergic cell groups C1-C3,
– serotoninergic cell groups B1-B8.

Mossy fibers

Cerebellum 2

Mossy fibers connect motor centers from all parts of the central nervous system with the cerebellum and they are elements of all important afferent bundles of the cerebellum.

Mossy fibers terminate bilaterally in discrete areas of the vermis and of the hemispheres, evidencing a partial somatotopic arrangement.

Motor cells

Medulla spinalis 1

Motoneurons situated in the anterior horn of the spinal cord. Their axons project without further synapsing to the peripheral skeletal muscles.

Motor cortex, primary

Telencephalon
TA Latin: *Gyrus precentralis*
TA English: *Precentral gyrus*
→ Precentral gyrus (area 4)

Motor cortex, supplementary

Telencephalon 2

Together with the premotor cortex, forms the secondary motor field. Involved in planning movement.

Motor nuclei 1

TA Latin: *Nuclei motorii*
TA English: *Motor nuclei*
Motor nuclei of cranial nerves V, VII, IX and X. Cranial nerves IX and X have their somatomotor roots in nucleus ambiguus.

Motor nucleus of the facial nerve

TA Latin: *Nucl. n. facialis*
TA English: *Motor nucleus of facial nerve*
→ Nucleus of facial nerve

Motor nucleus of the trigeminal nerve

Mesencephalon 3

TA Latin: *Nucl. motorius n. trigemini*
TA English: *Motor nucleus of trigeminal nerve*
The motor nucleus of the trigeminal nerve lies directly beside the principal nucleus of the trigeminal nerve. It controls masticatory musculature. Its efferents run behind the trigeminal ganglion in the mandibular nerve (V3) and, having

entered the oval foramen, they divide into four branches: the masseter nerve, deep temporal nerves, pterygoid nerves and mylohyoid nerve.

Motor root of the trigeminal nerve (proprioceptive fibers)

Nerves 3

TA Latin: *N. trigeminus, radix motoria*
TA English: *Motor root of trigeminal nerve*
Although not suggested by the name, the motor root of the trigeminal nerve (V) also contains sensory fibers, the proprioceptive fibers. These are muscle-spindle afferents which are the sole primary afferents of the body having their perikarya within the CNS itself. These perikarya form the mesencephalic nucleus of the trigeminal nerve

Motor root of trigeminal nerve (V)

Nerves 3

TA Latin: *N. trigeminus (N.V), radix motoria*
TA English: *Motor root of trigeminal nerve (V)*
The motor portions of the trigeminal nerve (V) originate in the motor nucleus of the trigeminal nerve. Together with other portions, they go to the ciliary ganglion, passing it without synapsing, and then coursing to the mandibular nerve (V3), the largest trigeminal branch, through the oval foramen into the intratemporal fossa. Four motor nerves arise here: the masseter nerve, deep temporal nerves, pterygoid nerves and mylohyoid nerve.

Myelencephalon

Myelencephalon 3

TA Latin: *Myelencephalon*
TA English: *Myelencephalon*
Syn: Medulla oblongata. The lowest portion of the brainstem. Here in the transition region to the spinal cord are important nuclear regions (olive, pyramid, gracile nucleus, cuneate nucleus), the cranial nerve nuclei of nerves V–XII as well as the centers of respiratory control. Important pathways are the medial lemniscus (somatosensory), lateral lemniscus (auditory) and vestibulospinal tract.

N

Neocerebellum

Cerebellum   2

TA Latin: *Neocerebellum*

TA English: *Neocerebellum*

Phylogenetically, a very young part of the cerebellum. The neocerebellum contains the two cerebellar hemispheres and receives afferents mainly from the pons, thus being called also the pontocerebellum.

Neocortex

Telencephalon   3

TA Latin: *Neocortex*

TA English: *Neocortex*

The cerebral cortex consists of two types of tissue:

- allocortex: 3- to 5-layer tissue without a uniform pattern. Frequently encountered in the olfactory system and in the limbic system.
- cerebral cortex: 6-layer nervous tissue accounting for around 95% of the cortex and endowed with a thickness of between 2 mm (visual system) and 5 mm (motor cortex).

Neuraxis

General CNS   3

TA Latin: *Neuraxis*

TA English: *Neuraxis*

Central nervous system. Composed of the encephalon located in the skull and the spinal cord running in the vertebral canal.

Neurohypophysis

Diencephalon   3

TA Latin: *Neurohypophysis*

TA English: *Neurohypophysis*

→ Posterior lobe of the hypophysis

Nigrostriatal fibers

Pathways   2

Somatotopic, dopaminergic projection of the substantia nigra, pars compacta, to the corpus striatum.

Nodulus

Cerebellum   3

TA Latin: *Nodulus*

TA English: *Nodule*

The vermis segment nodulus and the hemisphere segment flocculus together form the flocculonodular lobe.

Phylogenetically it is very old and is thus called the archicerebellum. Since its afferents come mainly from the vestibular nuclei (vestibulocerebellar tract), the "vestibulocerebellum" is another synonym.

Noradrenergic cell group A3   1

It has been possible to identify this noradrenergic cell group only in rats but not in primates.

Noradrenergic cell groups A1-A7

General CNS   2

TA Latin: *Cellulae noradrenergicae (A1–A7)*

TA English: *Noradrenergic cell s (A1–A7)*

There are seven different cell groups using noradrenaline as a transmitter and denoted by the letter "A".

Together they form the "noradrenergic group" of the lateral reticular formation, thus belonging to the monoaminergic cells.

The cell clusters are generally associated with other nuclei, with around the half of all noradrenergic neurons being located in the locus coeruleus (A6).

Nuclei of posterior column

TA Latin: *Nuclei columnae post.*

TA English: *Nuclei of posterior column*

In the cuneate nucleus and gracile nucleus terminate the epicritic afferents of the posterior column – funiculus dorsalis – (cuneate fasciculus and gracile fasciculus), which is the reason why they are also called posterior column nuclei.

- gracile nucleus: afferents from the trunk and lower extremities.

– cuneate nucleus: afferents from the upper extremities and neck (medial cuneate nucleus) and vestibular organ (lateral cuneate nucleus).

The efferents of both nuclei cross to the contralateral side in the medulla as the internal arcuate fibers and join the trigeminal efferents (epicritic sensibility of the face) to form the medial lemniscus, before passing to the thalamus (ventral posterolateral thalamic nucleus), from where they project into the somatosensory cortex (postcentral gyrus).

Nucleus ambiguus

Myelencephalon   3
TA Latin: *Nucl. ambiguus*
TA English: *Nucl. ambiguus*
Like the dorsal nucleus of the vagus nerve and the nucleus of the hypoglossal nerve, the ambiguus features a cellular column of at least 2 cm in length and also runs parallel to these nuclei. This is no surprise as it is the origin of somatomotor (actually special visceromotor) fibers of glossopharyngeal nerve (IX) and vagus nerve (X), which are responsible for innervation of the pharynx and larynx muscles.

Nucleus of abducens nerve   3

TA Latin: *Nucl. n. abducentis*
TA English: *Nucleus of abducens nerve*
The nucleus of the abducens nerve (VI) is enveloped by the genu of the facial nerve and, in addition to the actual motoneurons for the lateral rectus muscle of eye, it contains an almost equally great number of small neurons which are responsible for eye movements.

Nucleus of facial nerve

Pons   3
TA Latin: *Nucl. n. facialis*
TA English: *Motor nucleus of facial nerve*
This motor nucleus of cranial nerve VII is almost 5 mm long and lies in the caudal extension of the motor nucleus of the trigeminal nerve.
Two parts can be identified:

– nucleus of the facial nerve, dorsal part, innervates forehead muscles and the orbicular muscle of eye.
– nucleus of the facial nerve, ventral part, innervates perioral musculature.

Afferents come from the cerebral cortex and the pontine reticular formation.

Nucleus of facial nerve, accessory cells

Pons   2
Cell groups around the actual nucleus of the facial nerve innervate the stapedius muscle and some other muscles (probably belly of digastric muscle, stylohyoid muscle etc.).

Nucleus of the hypoglossal nerve

Pons   3
TA Latin: *Nucl. n. hypoglossi*
TA English: *Nucleus of hypoglossal nerve*
This nucleus that measures just under 2 cm in length runs parallel to the dorsal nucleus of the vagus nerve, in the lower angle of the fourth ventricle, which features a similarly named area (hypoglossal trigone). Afferents come via the contralateral corticobulbar tract from the cerebral cortex as well as from the pontine regions of the lateral reticular formation.

Nucleus of the lateral mammillary body

TA Latin: *Nucl. mammillaris lat.*
TA English: *Lateral nucleus of mammillary body*
→ Mammillary body, lateral nucleus

Nucleus of the oculomotor nerve

Mesencephalon   3
TA Latin: *Nucl. n. oculomotorii*
TA English: *Nucleus of oculomotor nerve*
The oculomotor nerve has its roots in a cluster of cells between the superior colliculus from which two groups can be isolated: the large-celled part is responsible for somatomotor control, while the small-celled part attends to visceromotor control of the eye.
The small-celled part is also called the accessory nucleus of the oculomotor nerve, the large-celled is composed of five groups, evidencing somatotopic arrangement as regards eye muscles.

Nucleus of the spinal root of the accessory nerve

Medulla spinalis   3
→ Accessory nerve (XI)

Nucleus of the superior lateral olive

Mesencephalon   2
TA Latin: *Nucl. olivaris sup. lat*
TA English: *Lateral superior olivary nucleus*
Lateral portion of the superior olive.
Projects to the ipsilateral inferior colliculus,
centralis nucleus and exerts influence on the
tensor tympani muscle and stapedius muscle via
the motor nucleus of the trigeminal nerve and
the nucleus of the facial nerve.

Nucleus of the superior olive

Mesencephalon   2
TA Latin: *Nucl. olivaris sup.*
TA English: *Superior olivary nucleus*
→ Superior olive

Nucleus of the trapezoid body

Mesencephalon   2
TA Latin: *Nucl. corporis trapezoidei*
TA English: *Nucleus of trapezoid body*
Belongs to the superior olivary complex and is
thus part of the ascending auditory projection.
Afferents come from the ventral cochlear nucleus. Efferents pass on to the nucleus of the superior lateral olive, from which they then go to
the inferior colliculus.

Nucleus of the trochlear nerve

Mesencephalon   3
TA Latin: *Nucl. n. trochlearis*
TA English: *Nucleus of trochlear nerve*
The nuclear region of the trochlear nerve (IV)
lies caudal to the oculomotor nucleus, between
the two inferior colliculi.
Afferents are received from the prepositus
hypoglossal nucleus and the interstitial nucleus
(Cajal). Its afferents descend along the superior
medullary velum, crossing to the contralateral
side (decussation of the trochlear nerves) and
emerge immediately beneath the inferior colliculus from the brainstem.

Nucleus proprius

Medulla spinalis   3
TA Latin: *Nucl. proprius*
TA English: *Nucleus proprius*
Nucleus in the middle of the posterior horn of
the spinal cord. Present in all spinal cord seg-
ments, it is a synaptic center for proprioceptive

afferents from the locomotor apparatus. The
impulses pass on to the cerebellum via the anterior spinocerebellar tract where they are compared with setpoint values.

O

cesses are already taking place in area 18 and area 19.

Damage to the visual cortex of one hemisphere leads to anything from disruption of fields of vision (scotoma), directly correlated with the extent of damage, to homonymous hemianopsia (semi-blindness with disruption of one eye field).

If both visual cortices are affected, cortical blindness results. Eye reflexes such as pupillary reflex are preserved, but the cortex-related accommodation reflex is lost.

Obex   2
TA Latin: *Obex*
TA English: *Obex*
Lower tip of fourth ventricle.

Occipital artery
Vessels   3
TA Latin: *A. occipitalis*
TA English: *Occipital artery*
Arises from the external carotid artery and passes to the occipital region where it anastomoses via the parietal foramen with the middle meningeal artery, parietal branch.

At the level of the mastoid, it also gives off the mastoid branch which anastomoses via the mastoid foramen with the posterior meningeal artery.

Together they form the so-called extracraniomeningeal anastomoses.

Occipital artery, mastoid branch
Vessels   2
TA Latin: *A. occipitalis, R. mastoideus*
TA English: *Occipital artery, mastoid branch*
At the level of the mastoid, the mastoid branch is given off from the occipital artery and anastomoses via the mastoid foramen with the posterior meningeal artery.

Occipital gyri
Telencephalon   2
TA Latin: *Gyri occipitales*
TA English: *Occipital gyri*
Gyri of the occipital lobe, evidencing a varying shape.

At the calcarine sulcus are located the area 17 (striate cortex), in the surrounding gyri (area 18 and area 19) are situated the secondary visual cortex. Whereas area 17 is involved in direct processing of signals from the eye, cognitive pro-

Occipital lobe
Telencephalon   3
TA Latin: *Lobus occipitalis*
TA English: *Occipital lobe*
Extends from the occipital pole to the parieto-occipital sulcus.

Occipital pole
Telencephalon   2
TA Latin: *Polus occipitalis*
TA English: *Occipital pole*
Occipital pole of the brain located on the occipital lobe.

Occipital sinus
Vessels   3
TA Latin: *Sinus occipitalis*
TA English: *Occipital sinus*
The occipital sinus arises from the confluence of the sinuses and runs downwards along the falx cerebri. At the great occipital foramen, it divides into the marginal sinuses that transport venous blood to the superior bulb of the jugular vein.

Occipital sulcus
Telencephalon   3
The occipital sulcus stretches across the medial aspect of the occipital lobe. Around it lie area 17, the striate cortex.

Occipital vein
Vessels   2
TA Latin: *V. occipitalis*
TA English: *Occipital vein*
Collects venous blood from the cranial meninges (via occipital emissary vein) and superficial soft tissues of the occipital region and

carries it into the deep cervical vein and the vertebral venous plexus.

Occipital veins
Vessels
TA Latin: *Vv. occipitales*
TA English: *Occipital veins*
Frontal veins, frontoparietal veins, parietal veins and occipital veins form together the superior cerebral veins.
All these veins carry venous blood from their respective catchment areas to the superior sagittal sinus.

Occipitopontine tract
Pathways   2
Projections of the occipital lobe to the nuclei of the pons (Varolius).

Occipitotemporal sulcus
Telencephalon   2
TA Latin: *Sulcus occipitotemporalis*
TA English: *Occipitotemporal sulcus*
Sulcus between occipital gyri and lateral occipitotemporal gyrus.
Joins the pre-occipital notch, separating occipital lobe and temporal lobe.

Octavian nerve
Nerves
TA Latin: *N. vestibulocochlearis*
TA English: *Vestibulocochlear nerve*
→ Vestibulocochlear nerve (VIII)

Octavian nerve, cochlear part
Nerves
TA Latin: *N. cochlearis*
TA English: *Cochlear nerve*
→ Cochlear nerve

Octavian nerve, vestibular part
Nerves
TA Latin: *N. vestibularis*
TA English: *Vestibular nerve*
→ Vestibular nerve

Ocular dominance columns
Telencephalon
In Area 17 (primary visual cortex) the neurons are summed up in pillar-like function units, located vertical to the cortex surface, which proceed through all layers and are innervated by the right and left eye respectively. These function units are called ocular dominance pillars or columns.

Oculomotor nerve (III)
Nerves   3
TA Latin: *N. oculomotorius (N.III)*
TA English: *Oculomotor nerve (III)*
The oculomotor nerve is a motor cranial nerve endowed with both somato- and visceromotor components, for which one complex is responsible in each case. Together with the trochlear nerve (IV) and abducens nerve (VI) it controls eye movements.
It is involved in the lateral and medial eyeball movements (lateral rectus muscle and superior oblique muscle), raising of the palpebra as well as accommodation (ciliary muscle) and adaptation (sphincter muscle of pupil).
Skull: superior orbital fissure.

Olfactory bulb
Telencephalon   3
TA Latin: *Bulbus olfactorius*
TA English: *Olfactory bulb*
= Primary olfactory area.
The olfactory bulb is an evagination of the mesencephalon. The cytoarchitecture evidences several cell layers, akin to the retina, which are involved in processing of signals from the bipolar sensory cells of the olfactory epithelium. The processing results are passed on via the olfactory tract to the secondary olfactory cortex (at the level of the anterior perforated substance and ambiens gyrus).

Olfactory epithelium
Nose   2
In the upper section of the nasal cavity, the olfactory region (olfactory epithelium), measuring a few cm^2 in size. Here is found the olfactory epithelium composed of sensory, microvilli, support and basal cells. At their dendritic extensions, the bipolar sensory cells have clublike en-

largements bearing kinocillia, with which they enter the nasal cavity. With their basal processes, they stretch as far as the olfactory bulb.

Olfactory nerve (I)

Nerves   3
TA Latin: *N. olfactorius (N.I)*
TA English: *Olfactory nerve (I)*
The olfactory nerve is the term used for the sum of all of the very fine fibers passing from the bipolar (primary) sensory cells of the olfactory epithelium through the cribiform plate to the primary olfactory area, the olfactory bulb, and are called olfactory nerve (I). There they converge in a ratio of 1:100 on the cells of the bulb.
Skull: cribriform plate.
If the olfactory nerve (I) is damaged, e.g. as a sequel of a base of skull injury, depending on the extent of injury, hyposmia or anosmia can ensue. Pungent substances such as ammonia can still be smelt as the nasal mucosa are stimulated, stimulating in turn the trigeminal nerve.

Olfactory sulcus

Telencephalon   2
TA Latin: *Sulcus olfactorius*
TA English: *Olfactory sulcus*
Longitudinal sulcus on the basal surface of the frontal lobe. The olfactory tract runs here.

Olfactory system

Telencephalon   3
→ Rhinencephalon

Olfactory tract

Telencephalon   3
TA Latin: *Tractus olfactorius*
TA English: *Olfactory tract*
The olfactory tract conducts the nerve impulses from the olfactory tract (primary olfactory area) to the olfactory cortex (secondary olfactory area). The latter contains the anterior olfactory nucleus, the olfactory tuberculum, the prepiriform cortex, the ambiens gyrus as well as cortical amygdaloid nucleus.
At its end, the tract divides into the lateral olfactory stria and medial olfactory stria.

Olfactory trigone

Telencephalon   2
TA Latin: *Trigonum olfactorium*
TA English: *Olfactory trigone*
Triangular area on the anterior perforated substance, formed by the two extensions of the olfactory bulb, i.e. the lateral olfactory stria and the medial olfactory stria.

Olfactory tuberculum

Telencephalon   2
TA Latin: *Tuberculum olfactorium*
TA English: *Olfactory tubercle*
The olfactory tuberculum is very poorly developed in humans and consists of a thin cell layer lying directly in the olfactory trigone. It receives fibers directly from the olfactory bulb.

Olfactory vein

Vessels
TA Latin: *V. gyri olfactorii*
TA English: *Vein of olfactory gyrus*
→ Vein of the olfactory gyrus

Olive   3

TA Latin: *Oliva*
TA English: *Olive*
– inferior olive is the actual "olive" and is located directly beneath the pons, in the myelencephalon. This large nucleus plays a major role in movement coordination.
– superior olive: nuclear conglomeration in the mesencephalon, is a component of the auditory tract.

Olivocerebellar fibers

Cerebellum/Pathways   2
TA Latin: *Tractus olivocerebellaris*
TA English: *Olivocerebellar tract*
Fibers of the olivocerebellar tract.
The latter conducts information from the inferior olivary complex to the cerebellar hemispheres. It is an important component of major functional and feedback loops which controls pending fine motor movements programs.
→ Olivocerebellar tract

Olivocerebellar tract
Cerebellum/Pathways   3
TA Latin: *Tractus olivocerebellaris*
TA English: *Olivocerebellar tract*
Large fiber bundle connecting the olive with the cerebellum.

The fibers emerge form the inferior olivary complex at the hilum of the olive, decussate in the contralateral side at the intermediate layer of the olive and enter the cerebellum via the inferior cerebellar peduncle. There they give off collaterals to the cerebellar nuclei before reaching the dendrites of the Purkinje cells as climbing fibers.

Olivocochlear bundle
Pathways
TA Latin: *Tractus olivocochlearis (Rasmussen)*
TA English: *Olivocochlear tract (Rasmussen)*
→ Olivocochlear fasciculus (Rasmussen)

Olivocochlear fasciculus (Rasmussen)
Pathways   2
TA Latin: *Tractus olivocochlearis (Rasmussen)*
TA English: *Olivocochlear tract (Rasmussen)*
The olivocochlear fasciculus or the Rasmussen fasciculus originates in the peri-olivary nuclei and is composed largely of fibers which decussate in the median plane and course through the reticular formation to the cochlear nuclei on the contralateral side.

Olivocochlear tract
Pathways
TA Latin: *Tractus olivocochlearis (Rasmussen)*
TA English: *Olivocochlear tract (Rasmussen)*
→ Olivocochlear fasciculus (Rasmussen)

Operculofrontal artery
Vessels
TA Latin: *A.prefrontalis*
TA English: *Prefrontal artery*
→ Prefrontal artery (candelabra artery)

Ophthalmic artery
Vessels   3
TA Latin: *A. ophthalmica*
TA English: *Ophthalmic artery*

The ophthalmic artery arises from the internal carotid artery, passes through the optic canal to the orbita, supplying here the walls of the orbita, the eye, eye muscles, lacrimal apparatus, palpebra superior, frontal sinus and nose. Cranial meninges are likewise supplied.

Ophthalmic branch of the trigeminal nerve
Nerves
TA Latin: *R. ophthalmicus n. trigemini*
TA English: *Ophthalmic branch of trigeminal nerve*
→ Ophthalmic nerve (V1)

Ophthalmic nerve (V1)
Nerves   3
TA Latin: *N. ophthalmicus (N.V1)*
TA English: *Ophthalmic nerve (V1)*
This purely sensory branch goes to the orbita where it divides into three branches: nasociliary nerve, frontal nerve and lacrimal nerve.
Sensory innervation is effected via these nerves for:
– the entire orbita,
– the cornea,
– the forehead,
– the paranasal sinus,
– nasal septum.
Skull: superior orbital fissure.

Optic canal
Skeleton   3
TA Latin: *Canalis opticus*
TA English: *Optic canal*
The optic canal connects orbita to cranial fossa. The optic nerve passes through it.

Optic chiasm
Diencephalon   3
TA Latin: *Chiasma opticum*
TA English: *Optic chiasm*
Having entered via the optic canal, the two optic nerves converge and meet above the hypophysis, forming the optic chiasm.
Here the fibers of the respective retinal halves (=lateral field of vision) cross to the contralateral side, while the fibers of the lateral retina (= medial field of vision) continue further uncrossed. They continue as the optic tract to the

LGB where they synapse. LGB fibers project to the visual cortex.

Optic layer of the superior colliculus
Mesencephalon
TA Latin: *Stratum opticum colliculi sup.*
TA English: *Optic layer of superior colliculus*
→ Superior colliculus, optic layer

Optic nerve (II)
Nerves   3
TA Latin: *N. opticus (N.II)*
TA English: *Optic nerve (II)*
The optic nerve is purely a sensory nerve which is composed of axons from the retinal ganglia cells and passes as far as the optic chiasm. Here the fibers divide again (no synapse), to then continue as the optic tract to the lateral geniculate body.
Conveyed are the results from processing of visual stimuli in the cell layers of the retina.
Skull: optic canal.

Optic radiation
Nerves   3
TA Latin: *Radiatio optica*
TA English: *Optic radiation*
The visual radiation is the term used to designate the ray-shaped fiber bundles that leave the lateral geniculate body and at the lateral wall of the lateral ventricle pass on to the area 17 (striate cortex) at the occipital pole.
They conduct the visual raw material after being processed by the LGB.
Also called geniculocalcarine tract.

Optic radiation, occipital genu
Nerves   1
The optic radiation stretches from the LGB to the roof of the lateral ventricle, descending from there and bending caudally, so as to track the course of the side wall of the lateral ventricle. This curvature is called the occipital genu.
The temporal genu is situated on the end of the ventricle, at the so-called lateral ventricle, posterior horn. Around this, fiber components bend downwards and then upwards also producing a knee-shaped form.

Optic radiation, temporal genu
Nerves
→ Optic radiation, occipital genu

Optic recess
Meninges & Cisterns   1
Extension from the fourth ventricle to the roof of the optic chiasm. The infundibular recess begins beneath the chiasm.

Optic tract
Nerves/Pathways   3
TA Latin: *Tractus opticus*
TA English: *Optic tract*
The optic tract begins after the optic chiasm (there is no synaptic interchange).
It courses in the direction of the thalamus, dividing shortly before arrival into two parts: optic tract, lateral root goes to the lateral geniculate body, while the optic tract, medial root goes to the tegmentum of mesencephalon (superior colliculi) and to the pretectal region (terminal nuclei).

Optic tract, lateral root
Nerves   3
TA Latin: *Tractus opticus, radix lat.*
TA English: *Optic tract, lateral root*
Shortly before reaching the LGB, the optic tract divides:
1) lateral root goes to the LGB
2) medial root goes to the superior colliculus of the quadrigeminal plate, with various lateral branches being given off to the preoptic area (accessory optic tract).

Optic tract, medial root
Nerves   3
TA Latin: *Tractus opticus, radix med.*
TA English: *Optic tract, medial root*
→ Optic tract, lateral root

Optic tract nucleus
Diencephalon   2
TA Latin: *Nucl. tractus optici*
TA English: *Nucleus of optic tract*
The optic tract nucleus lies in the mesencephalon near the superior colliculus. The nucleus is fused with the dorsal terminal nucleus and is

an important center of the subcortical pathway which mediates horizontal optokinetic nystagmus.

Opticofacial winking reflex
→ Superior colliculus

Orbital gyri
Telencephalon 2
TA Latin: *Gyri orbitales*
TA English: *Orbital gyri*
On the roof of the orbita are gyri of the frontal lobe. Belongs to the prefrontal cortex, which purportedly plays a major role in higher, associative and complex mental functions.

Orbital sulci
Telencephalon 2
TA Latin: *Sulci orbitales*
TA English: *Orbital sulci*
Inconstant groove running on the basal surface of the frontal lobe. The orbital gyri and sulci lie on the roof of the orbita.

Orbital veins
Vessels 1
TA Latin: *Vv. orbitae*
TA English: *Orbital veins*
Veins coursing in the inner angle of eye, forming an anastomosis between the angular vein and cavernous sinus. They collect venous blood from the eye muscles, eyeball, soft tissue of the orbita and cranial meninges.

Orbitofrontal cortex
Telencephalon 1
Part of the frontal lobe resting on the orbita. Is subdivided into two parts:

– centroposterior orbitofrontal cortex,
– lateroposterior orbitofrontal cortex.

Pronounced cholinergic innervation and involvement in the processing of olfactory stimuli.

Organum vasculosum of the lamina terminalis
Diencephalon 2
TA Latin: *Organum vasculosum laminae terminalis*

TA English: *Vascular organ of lamina terminalis*
Belongs to the circumventricular organs and is directly situated at the preoptic area. The OVLT is rich in receptors for angiotensin II and plays a role in water and electrolyte balance, blood pressure control and is also thought to be involved in controlling LHRH secretion. It contains afferents from the subfornical organ, hypothalamus and locus coeruleus. Efferents return to the hypothalamus.

Oval area
→ Oval nucleus

Oval fasciculus
Pons 1
→ Oval nucleus

Oval nucleus
Pons 1
The oval nucleus is a dorsal extension of the solitary nucleus. Ascending root fibers of the facial nerve terminate here. Efferents going in the direction of the thalamus course in the oval fasciculus.

Oxytocin
2
Hormon of the paraventricular nucleus of hypothalamus. Oxytocin effects contraction of the uterus when giving birth and controls the release of milk during the lactation phase.

P

Pacchionian granulations   2
TA Latin: *Granulationes arachnoideae*
TA English: *Arachnoid granulations*
→ Arachnoid granulations (Pacchioni)

Pachymeninx
Meninges & Cisterns   3
TA Latin: *Pachymeninx*
TA English: *Pachymeninx*
pachys (Greek) = tough, meninx
= Dura mater.

Palaeocerebellum
Cerebellum   3
TA Latin: *Paleocerebellum*
TA English: *Paleocerebellum*
Phylogenetically, a very old part of the cerebellum. Corresponds to the vermis cerebelli with its surrounding intermediate part (paravermal part). The afferents of this region come from the spinal cord, hence this part is also called the spinocerebellum.

Palatine vein
Vessels   2
Forms an anastomosis between the pterygoid plexus and facial vein. It runs in the side wall of the pharynx and collects venous blood from the soft palate and pharynx.

Pallidoreticular tract
Pathways   2
Descending fibers from the globus pallidus of the basal ganglia to the reticular formation.

Pallidotegmental fibers
Pathways
→ Pallidoreticular tract

Pallidum
Diencephalon   3
TA Latin: *Pallidum*
TA English: *Pallidum*
→ Globus pallidus

Papez neuronal circuit
General CNS   3
The mammillothalamic fasciculus, Vicq d'Azyr bundle conducts efferents of the mammillary body to the thalamus (anterior thalamic nucleus). This in turn projects via the cingulum to the hipppocampus, while the latter projects back via the fornix to the mammillary body and anterior thalamic nucleus.
This creates a neuronal feedback circuit, which is called the Papez neuronal circuit and plays a role in memory formation.
Being a vital component of the Papez neuronal circuit, the hippocampus is involved in memory formation. Lesions result in loss of the ability transfer the contents from short-term memory to long-term memory (anterograde amnesia).

Papillioform nucleus
TA Latin: *Nucl. reticularis tegmenti pontis*
TA English: *Reticulotegmental nucleus*
→ Tegmental pontine reticular nucleus (Bechterew)

Parabigeminal body
Mesencephalon   2
TA Latin: *Nucl. parabigeminalis*
TA English: *Parabigeminal nucleus*
Small nucleus beside the inferior colliculus. Receives afferents from the ipsilateral superior colliculus and projects to the superior colliculi of both sides.

Parabrachial area
Diencephalon   2
TA Latin: *Nuclei parabrachiales*
TA English: *Parabrachial nuclei*
The parabrachial area comprises three nuclear areas:
– lateral parabrachial nucleus,
– medial parabrachial nucleus,
– Kölliker-Fuse nucleus.
The nuclei are involved in the processing of gustatory signals and breathing.

Parabrachial nuclei
Diencephalon 2
TA Latin: *Nuclei parabrachiales*
TA English: *Parabrachial nuclei*
The parabrachial nuclei lie together with the Kölliker-Fuse nucleus in the parabrachial area. There are two nuclei:
– lateral parabrachial nucleus,
– medial parabrachial nucleus.
Both nuclei are involved in processing of gustatory and similar signals from the respiratory tract.

Paracentral artery
Vessels 2
TA Latin: *A. pericallosa, R. paracentralis*
TA English: *Pericallosal artery, paracentral branch*
Together with the precuneal artery, it arises from the internal parietal artery and ascends to the margin of the hemisphere. Supplies the paracentral lobule.
Together with the posterior branch, the internal parietal artery emerges from the pericallosal branch.

Paracentral lobule
Telencephalon 3
TA Latin: *Lobulus paracentralis*
TA English: *Paracentral lobule*
A small lobe (lobule) descending deeply into the longitudinal fissure of cerebrum and belonging likewise to the premotor cortex.

Paracentral nucleus
TA Latin: *Nucl. paracentralis thalami*
TA English: *Paracentral nucleus of thalamus*
The intralaminar nuclei of the thalamus can be divided in two groups:
1) the caudal group (centromedia nucleus and parafascicular nucleus) and
2) the rostral group (central lateral nucleus, paracentral nucleus and central medial nucleus).
The nuclei play an important role in the perception of pain and they project to cortex and striatum.

Parafascicular + centromedial nuclei
Diencephalon 1
TA Latin: *Nuclei parafascicularis + centromedianus*
TA English: *Parafascicular + centromedian nuclei*
Both nuclei form the caudal group of the intralaminar nuclei of the thalamus. Both receive afferents from the globus pallidus and both project to the corpus striatum, with the centromedian nucleus projecting exclusively to the putamen, and the parafascicular nucleus only to the caudate nucleus.
Both nuclei are involved in the integration of polysensory information to the corpus striatum.

Parafascicular thalamic nucleus
Diencephalon 2
TA Latin: *Nucl. parafascicularis thalami*
TA English: *Parafascicular nucleus of thalamus*
The parafascicular nucleus belongs to the intralaminar nuclei of the thalamus and receives pronounced afferents from the cerebellum (fastigial nucleus and dentate nucleus). As is typical in the case of intralaminar nuclei, projections reach the striatum and frontal lobe/parietal lobe. The parafascicular nucleus projects exclusively to the caudate nucleus of the corpus striatum. The amygdaloid body also receives fibers.

Parahippocampal gyrus
Telencephalon 3
TA Latin: *Gyrus parahippocampalis*
TA English: *Parahippocampal gyrus*
The gyrus marks the transition from hippocampus with its allocortex to the isocortical structure of the temporal lobe. A cross-section shows 4 discrete cortical regions: presubiculum and parasubiculum on the hippocampal sulcus, entorhinal area and the perirhinal cortex deep in the calcarine sulcus.

Paramedian pontine medial reticular formation
Pons 1
Nuclei of the medial reticular formation, at the level of the pons, near the median centers (raphe nuclei).

Paramedian reticular nucleus
TA Latin: *Nucl. reticularis paramedianus*
TA English: *Paramedian reticular nucleus*
→ Anterior funicle nucleus

Parasubiculum
Telencephalon   1
TA Latin: *Parasubiculum*
TA English: *Parasubiculum*
Parasubiculum and presubiculum belong to the hippocampal region and evidence a (inhomogeneous) 4-layered structure.
Afferents come from the olfactory system, dorsal thalamus, as well as the contralateral presubiculum.
Efferents go to the entorhinal area, hippocampus, mammillary body and tegmentum of mesencephalon.

Parasympathetic nucleus of the oculomotor nerve
TA Latin: *Nucl. accessorius n. oculomotorii*
TA English: *Accessory nucleus of oculomotor nerve*
→ Accessory nucleus of oculomotor nerve

Parasympathetic nucleus of the vagus nerve
TA Latin: *Nucl. post. n. vagi*
TA English: *Posterior nucleus of vagus nerve*
→ Dorsal nucleus of the vagus nerve

Paraterminal gyrus
Telencephalon   2
TA Latin: *Gyrus paraterminalis*
TA English: *Paraterminal gyrus*
A gyrus situated on the median surface of the frontal lobe, directly rostral to the anterior commissure.
The induseum griseum running on the roof of the corpus callosum and around the genu of the corpus callosum joins the paraterminal gyrus. From here it continues further via the diagonal band (Broca) to the anterior perforated substance.

Paraventricular hypothalamic nucleus
Diencephalon   3
TA Latin:
Nucl. paraventricularis hypothalami

TA English: *Paraventricular nucleus of hypothalamus*
Two parts are identified: the magnocellular part produces vasopressin (antidiuretic) and oxytocin (contraction of uterus and mammary gland). The parvocellular part produces CRF (via ACTH influences motivation/emotion) and has myriad subcortical projections.
Afferent signals from a very large number of subcortical centers, as well as via hormones (e.g. sexual steroids).

Parietal foramen
Meninges & Cisterns   1
TA Latin: *Foramen parietale*
TA English: *Parietal foramen*
Inconstant point of passage of the occipital artery through the calvaria.

Parietal lobe
Telencephalon   3
TA Latin: *Lobus parietalis*
TA English: *Parietal lobe*
Extends from the central sulcus to the parieto-occipital sulcus.

Parietal operculum
Telencephalon   2
TA Latin: *Operculum parietale*
TA English: *Parietal operculum*
The part of the parietal lobe, which covers the insula (of Reil).

Parietal trunk of the middle cerebral artery
Vessels
→ Middle cerebral artery, parietal trunk

Parietal veins
Vessels
TA Latin: *Vv. parietales*
TA English: *Parietal veins*
Frontal veins, frontoparietal veins, parietal veins and occipital veins form together the superior cerebral veins.
All these veins carry venous blood from their respective catchment areas to the superior sagittal sinus.

Parieto-occipital artery

Vessels

Dysfunction of the artery of the angular gyrus produce a combination of aphasia, alexia and hemianopsia. Dysfunction of the supramarginal artery leads to hypoperfusion of the optic radiation, thus causing hemianopsia.

→ Artery of the angular gyrus

Parieto-occipital sulcus

Telencephalon   3

TA Latin: *Sulcus parietooccipitalis*

TA English: *Parieto-occipital sulcus*

The parieto-occipital sulcus is an extension of the calcarine sulcus. It rises from a deep level of the longitudinal fissure of cerebrum and terminates above the occipital gyrus.

Parieto-occipito-temporopontine tract

Pathways   2

Projections from the parietal lobe, occipital lobe and temporal lobe to the pontine nuclei.

Parietopontine tract

Pathways   2

Projections of mostly motor information from the parietal lobe to the pontine nuclei (Varolius).

Parietotemporopontine tract

Pathways   2

Projections of mostly motor information from the parietal lobe and temporal lobe to the pontine nuclei (Varolius).

Parkinson´s disease

The Parkinson's disease is characterized by progressive loss of the neurons in the black substance, compact part, the degeneration of its ascending projection and reduction of the dopamine contents in striate body. Symptoms are rigidity, tremor, akinesia.

Parvocellular reticular nucleus

Pons   2

TA Latin: *Nucl. reticularis parvocellularis*

TA English: *Parvocellular reticular nucleus*

Nucleus in the rhombencephalon belonging to the lateral reticular formation and situated di-

rectly medial to the sensory nucleus of the trigeminal nerve. It is involved in brainstem reflexes, and receives afferents from brain segments situated in higher regions and projects to motor nuclei of the tegmentum area.

Peduncle of lentiform nucleus

Telencephalon   2

The location at which the caudal caudate nucleus attaches to the lentiform nucleus is called the peduncle of lentiform nucleus.

Peduncle of mammillary body

Diencephalon   1

The bundle conveys fibers of the dorsal trigeminal nucleus (Gudden) to the mammillary body.

Peduncular branches

Vessels

TA Latin: *Rr. pedunculares a. cerebri post.*

TA English: *Peduncular branches of posterior cerebral artery*

→ Posterior cerebral artery, peduncular branches

Peduncular veins

Vessels   2

TA Latin: *Vv. pedunculares*

TA English: *Peduncular veins*

Inconstant lateral branches of the basal vein which drain the cerebral peduncles.

Pedunculopontine tegmental nucleus (Ch.5)   1

TA Latin: *Nucl. Tegmentalis pedunculopontinus (Ch.5)*

TA English: *Pedunculopontine tegmental nucleus (Ch.5)*

An important nucleus from the cholinergic cell group of the lateral reticular formation. It has two parts:

- pedunculopontine tegmental nucleus, compact part
- pedunculopontine tegmental nucleus, diffus part

Pedunculopontine tegmental nucleus, compact part

Mesencephalon   3

TA Latin: *Nucl. Tegmentalis*
pedunculopontinus, pars compacta
TA English: *Pedunculopontine tegmental nucleus, compact part*
This densely packed part of the pedunculopontine tegmental nucleus lies in the caudolateral mesencephalon and has reciprocal connections with the motor centers and the limbic system. Efferents go to the spinal cord.
Electrical stimulation of this area causes coordinated locomotion ("mesencephalic locomotor region") in decerebrated animals.

Pedunculopontine tegmental nucleus, diffuse part

Mesencephalon   1
TA Latin: *Nucl. Tegmentalis pedunculopontinus, pars dissipata*
TA English: *Pedunculopontine tegmental nucleus, dissipated part*
In addition to the pedunculopontine tegmental nucleus, compact part, the pedunculopontine tegmental nucleus also contains a cholinergic region with loosely arranged cell bodies. But their function is not clear unlike that of the locomotor tasks of the compact part.

Periamygdaloid cortex

Telencephalon   1
TA Latin: *Cortex periamygdaloideus*
TA English: *Periamygdaloid cortex*
Cortex area around the amygdaloid body. Projects to the central amygdaloid nucleus.

Periaqueductal gray matter

TA Latin: *Substantia grisea centralis*
TA English: *Periaqueductal gray substance*
→ Central gray matter of mesencephalon

Periaqueductal gray substance of metencephalon

→ Central gray matter of mesencephalon

Pericallosal artery

Vessels   3
TA Latin: *A. pericallosa*
TA English: *Pericallosal artery*
Together with the callosomarginal artery, the pericallosal artery emerges from the anterior ce-

rebral artery, which in turn emerges from the internal carotid artery.
It passes around the genu of the corpus callosum and divides on the tectum of the corpus callosum into the posterior branch and the internal parietal artery, which shortly afterwards divides into the paracentral artery, precuneal artery and occipitotemporal artery.
Supplies the forebrain with its branches.

Pericallosal artery, posterior branch

Vessels   2
TA Latin: *A. pericallosa, R. parietooccipitalis*
TA English: *Pericallosal artery, parieto-occipital branch*
The pericallosal artery divides into the internal parietal artery and the posterior branch. The latter passes by the corpus callosum in the direction of the splenium of the corpus callosum where it anastomoses with the dorsal branch of the cingulothalamic artery.

Pericallosal cistern

Meninges & Cisterns   1
TA Latin: *Cisterna pericallosa*
TA English: *Pericallosal cistern*
The pericallosal cistern is formed on the roof of the corpus callosum, at the sulcus of corpus callosum. It stretches from the splenium of corpus callosum to the genu of the corpus callosum, where it joins the cistern of lamina terminalis. Coursing with it is the pericallosal artery.

Pericranium

Meninges & Cisterns   1
TA Latin: *Pericranium*
TA English: *Pericranium*
Pericranium is an abbreviation of periosteum cranii, i.e. the periosteum of the skull.
The pericranium is the entire periosteum enclosing the skull.

Peridural anesthesia

→ Epidural cavity

Periolivary nuclei

Mesencephalon   1
TA Latin: *Nuclei periolivares*
TA English: *Peri-olivary nuclei*

Diffuse collection of cells in the immediate vicinity of the superior olive.

They receive afferents from the contralateral ventral cochlear nucleus and project to the ipsilateral inferior colliculus.

Periosteum

Meninges & Cisterns 1

TA Latin: *Periosteum*

TA English: *Periosteum*

The periosteum is the entire skin enclosing a bone.

Peripeduncular nucleus

Mesencephalon 2

TA Latin: *Nucl. peripeduncularis*

TA English: *Peripeduncular nucleus*

The peripendicular nucleus can be viewed as being the caudalmost part of the zona incerta. It is bidirectionally connected with the ventromedial hypothalamic nucleus and is thought to project to the amygdaloid body and medial preoptic area and lateral hypothalamic area.

Perirhinal cortex

Telencephalon

→ Entorhinal + perirhinal cortices

Periventricular preoptic nucleus

Diencephalon

TA Latin: *Nucl. preopticus periventricularis*

TA English: *Periventricular preoptic nucleus*

→ Median preoptic nucleus

Periventricular thalamic nuclei

Diencephalon 1

TA Latin: *Nuclei periventriculares hypothalami*

TA English: *Periventricular hypothalamic nuclei*

The medial thalamic nucleus is called the periventricular nuclear region of the thalamus, since it forms a layer containing nerve cells, beneath the ventricle ependyma.

Pes hippocampi

Telencephalon 2

TA Latin: *Pes hippocampi*

TA English: *Pes hippocampi*

The approximately 5 cm long hippocampus in the floor of the lateral ventricle, inferior horn, terminates in a foot-shaped pattern, called pes hippocampi, while the individual evaginations are analogously called digitationes hippocampi (fingers).

Petrosal vein

Vessels 1

TA Latin: *V. petrosa*

TA English: *Petrosal vein*

The petrosal vein together with the lateral mesencephalic vein es lishes a connection between the basal vein and the petrosal sinus. In the process it receives a number of other vessels, e.g. superior hemisphere veins and transverse pontine vein.

Pia mater

Meninges & Cisterns 3

TA Latin: *Pia mater*

TA English: *Pia mater*

Together with the arachnoid, the pia mater forms the leptomeninx.

Whereas the arachnoid follows the course of the dura mater and hence of the calvaria, the pia mater rests on the surface of the brain, pursuing a joint course along the sulci.

Lying between the pia mater and arachnoid is the subarachnoid space that is filled with CSF.

Pigmental parabrachial nucleus

Diencephalon 1

This nuclear region lies in the dorsolateral segment of the tegmentum of mesencephalon and is part of the reticular formation. It contains efferents to the globus pallidus and, like the neighboring substantia nigra and the ventral tegmental area, it likewise produces dopamine.

Pineal body

Diencephalon 3

TA Latin: *Glandula pinealis*

TA English: *Pineal gland*

This nuclear region, shaped like a pine cone, above the quadrigeminal lamina forms, together with the habenular nuclei, the so-called epithalamus. The function of the pineal body, also called epiphysis or pineal gland, is to produce the hormone melatonin, which plays a role in light-controlled circadian rhythms. Afferents come from the suprachiasmatic nucleus, pregeniculate nucleus and posterior commissural nucleus.

Pineal recess

Meninges & Cisterns   2
TA Latin: *Recessus pinealis*
TA English: *Pineal recess*
Extension of the third ventricle reaching into the pineal body.

Polar plane

Telencephalon   2
The temporal plane, the upper side of the temporal lobe deep in the lateral sulcus, stretches to the temporal pole, the frontmost tip of the temporal lobe. This is called the polar plane.

Polyuria

→ Supraoptic nucleus

POML

TA Latin: *Nucl. preopticus med.*
TA English: *Medial preoptic nucleus*
→ Medial preoptic nucleus

POMN

TA Latin: *Nucl. preopticus medianus*
TA English: *Median preoptic nucleus*
→ Median preoptic nucleus

Pons (pontine nucleus)

Pons
TA Latin: *Pons (nuclei pontis)*
TA English: *Pons (pontine nuclei)*
→ Pontine nuclei

Pons (Varolius)

Pons   3
TA Latin: *Pons*
TA English: *Pons*
The pons consists of 2 parts: base of pons and tegmentum. The typical protruding base of pons accommodates the pyramidal tracts. Interspersed here are the pontine nuclei, where corticopontine fibers synapse. The tegmentum area contains cranial nerve nuclei (V,VI, VII,

VIII), trapezoid body, medial lemniscus, parts of the reticular formation and the medial longitudinal fasciculus.

Pontine arteries, medial branch

Vessels
TA Latin: *Aa. pontis, R. med.*
TA English: *Pontine arteries, medial branch*
→ Middle pontine arteries

Pontine "attack area"

Pons   1
A center located at the level of the pons that plays a decisive role in initiation of defense and attack behavior.

Pontine cistern (medial and lateral parts)

Meninges & Cisterns   1
The pontine cistern is situated before the cerebral pons, surrounding the fossa of the basilar sulcus of pons.

Pontine medial reticular formation

Pons   1
TA Latin: *Formatio reticularis tegmentum pontis*
TA English: *Pontine medial reticular formation*
Medial reticular formation at the level of the pons.

Pontine nuclei

Pons   3
TA Latin: *Nuclei pontis*
TA English: *Pontine nuclei*
Nuclear regions scattered loosely in the base of the pons, between the pyramidal tracts. Corticopontine fibers synapse here and the postsynaptic, olivocerebellar tract project to the contralateral cerebellar hemisphere. These fibers bestow on the pons their typical, horizontal pattern of stripes.

Pontine reticular formation

Pons   1
TA Latin: *Formatio reticularis pontis*
TA English: *Pontine reticular formation*
The pontine reticular formation is composed of two parts: caudal pontine reticular nucleus and oral pontine reticular formation. Efferents go

via the medial reticulospinal tract to the spinal cord. Both nuclei play a decisive role in controlling fast horizontal saccadic eye movements.

Pontine reticular nucleus
Pons 1
TA Latin: *Nucl. reticularis pontis*
TA English: *Pontine reticular nucleus*
→ Pontine reticular formation

Pontine veins
Vessels 2
TA Latin: *Vv. pontis*
TA English: *Pontine veins*
The pontine veins form a venous plexus around the pons. They collect venous blood from the pons and convey it to the superior petrosal vein and the anteromedial pontine vein.

Pontine venous plexus
Vessels
TA Latin: *Vv. pontis*
TA English: *Pontine veins*
→ Pontine veins

Pontobulbar body
Pons 1
TA Latin: *Nucl. pontobulbaris*
TA English: *Pontobulbar nucleus*
A nucleus of the same ontogenetic provenance as the pontine nucleus and which also performs the same function, i.e. synapsing of corticocerebellar fibers

Pontocerebellar cistern
Meninges & Cisterns 2
TA Latin: *Cisterna pontocerebellaris*
TA English: *Pontocerebellar cistern*
The pontocerebellar cistern surrounds the entire ventral side of the cerebellum, it runs from the middle cerebellar peduncle around the biventer lobule as far as the ventral extensions of the horizontal fissure.

Pontocerebellar fibers
Cerebellum 1
TA Latin: *Fibrae pontocerebellares*
TA English: *Pontocerebellar fibres*

Fibers of the pontocerebellar tract.
Its mossy fibers conduct information from the pontine nucleus directly to the cortical regions of the cerebellar hemispheres. It is an important component of major functional and feedback loops which regulate movement planning.

Pontocerebellum
Cerebellum 2
TA Latin: *Pontocerebellum*
TA English: *Pontocerebellum*
The hemispheres belong to the phylogenetic young neocerebellum and receive their afferences via the moss fibers of the pontocerebellar tract from the pontine nuclei. Therefore, one also likes to summarize all hemispheric sections to the so-called pontocerebellum.

Pontospinal fibers
Pathways
→ Pontospinal tract

Pontospinal pathway
Pathways
TA Latin: *Tractus pontoreticulospinalis*
TA English: *Pontoreticulospinal tract*
→ Pontospinal tract

Pontospinal tract
Pathways 2
TA Latin: *Tractus pontoreticulospinalis*
TA English: *Pontoreticulospinal tract*
The pontospinal tract or medial reticulospinal tract begins in the caudal pontine reticular nucleus and in the caudal portion of the oral pontine reticular nucleus. It descends mostly in the spinal cord in the medial portion of the anterior column. Its fibers terminate chiefly in laminae VII and VIII of the spinal gray matter; but they also pass in the part of lamina IX containing the motoneurons for the trunk musculature.

Postcentral artery
Vessels
TA Latin: *A. sulci postcentralis*
TA English: *Artery of postcentral sulcus*
→ Artery of postcentral sulcus
(anterior parietal artery)

Postcentral gyrus

Telencephalon   3

TA Latin: *Gyrus postcentralis*

TA English: *Postcentral gyrus*

= primary somatosensory cortex = SI

= area 3 + 1 + 2

The postcentral gyrus lies in the parietal lobe directly on the central sulcus. Observing strict somatotopic arrangement, the somatosensory tracts of the contralateral body half terminate here.

Conscious localization and differentiation of quality and intensity of a tactile stimulus are effected in cooperation with the postcentral gyrus. Lesions of the postcentral gyrus reduces the response to tactile, thermal and noci stimuli from the contralateral body half.

Postcentral sulcus

Telencephalon   3

TA Latin: *Sulcus postcentralis*

TA English: *Postcentral sulcus*

The postcentral sulcus extends in the parietal lobe almost parallel to the central sulcus and divides the primary somatosensory cortex from the secondary somatosensory cortex.

Postcommissural stria terminalis

Diencephalon   2

The stria terminalis is the most important efferent of the amygdaloid body and divides at the anterior commissure into three components:

1) the postcommissural stria terminalis runs to the interstitial nucleus of the stria terminalis,
2) the commissural fibers connect the two cortical nuclei of the amygdala,
3) the precommissural stria terminalis ends in the medial preoptic area, anterior hypothalamic area, hypothalamic nuclei.

Posterior and lateral choroid branches ; lateral posterior choroid artery

Vessels

TA Latin: *A. cerebri post., Rr. Choroidei posteriores lat.*

TA English: *Posterior cerebral artery, posterior lateral choroidal branches*

→ Posterior cerebral artery, lateral posterior choroid branch

Posterior and medial choroid branches; medial posterior choroid artery

Vessels

TA Latin: *A. cerebri post., Rr. Choroidei posteriores med.*

TA English: *Posterior cerebral artery, posterior medial choroidal branches*

→ Posterior cerebral artery, lateral posterior choroid branch

Posterior cerebellar lobe

Cerebellum   2

TA Latin: *Lobus cerebelli post.*

TA English: *Posterior lobe of cerebellum*

The posterior lobe is the part of the cerebellum caudal to the primary fissure, and is composed of vermis portions (declive, folium, tuber, pyramid and uvula) as well as hemisphere portions (simple lobule, semilunar, gracile and biventer lobules as well as tonsil). Functionally this subdivision has practically no significance, since the cerebellum evidences a functional arrangement in a vertical direction (vermis, intermediate part, lateral part).

Posterior cerebral artery

Vessels   3

TA Latin: *A. cerebri post.*

TA English: *Posterior cerebral artery*

Together with the anterior cerebral artery and middle cerebral artery, forms the three major cerebral vessels.

The right and left branches of the posterior cerebral artery emerge from the division of the basilar artery.

Supply area:

– subcortical: thalamus and parts of the midbrain
– cortical: parts of the temporal lobe, the entire occipital lobe (includ. visual cortex) as well as the hippocampus.

Posterior cerebral artery, anterior temporal branch

Vessels   1

TA Latin: *A. cerebri post., R. temporalis ant.*

TA English: *Posterior cerebral artery, anterior temporal branch*

Lateral branches of the posterior cerebral artery pass to the temporal lobes.

The lateral branches are heterogeneous in their expression and supply the basal regions of the temporal lobe.

Posterior cerebral artery, lateral posterior choroid branch

Vessels 2

TA Latin: *A. cerebri post., R. choroideus post. lat.*

TA English: *Posterior cerebral artery, posterior lateral choroidal branch*

These lateral branches arise from the postcommunical part (P2) of the posterior cerebral artery and supply the choroid plexus of the lateral ventricle and of the third ventricle.

Posterior cerebral artery, medial posterior choroid branch

Vessels 2

TA Latin: *A. cerebri post., R. choroideus post. med.*

TA English: *Posterior cerebral artery, posterior medial choroidal branch*

From the precommunical part of the posterior cerebral artery frequently arise small arteries coursing to the choroid plexus of the fourth ventricle and the third ventricle.

Posterior cerebral artery, peduncular branches

Vessels 1

TA Latin: *A. cerebri post., Rr. pedunculares*

TA English: *Posterior cerebral artery, peduncular branches*

Smaller arteries branch from the posterior cerebral artery as well as from other surrounding vessels into the tissue of the cerebral peduncles, supplying the latter.

Posterior cerebral artery, postcommunical part

Vessels 1

TA Latin: *A. cerebri post., pars Postcommunicalis*

TA English: *Posterior cerebral artery, postcommunicating part*

Also called P2 segment of the posterior cerebral artery. Extends from the posterior communicating artery to the posterior inferior temporal branch and courses along the mesencephalon. Situated in the cisterna ambiens.

Posterior cerebral artery, postcommunical part, thalamic branches

Vessels 1

TA Latin: *A. thalamogeniculata*

TA English: *Thalamogeniculate artery*

Lateral branches of the posterior cerebral artery which enter and supply the thalamus. They make provision for blood supply to the geniculate bodies, occipital limb of the internal capsule, as well as contiguous thalamic sections.

Posterior cerebral artery, precommunical part

Vessels 3

TA Latin: *A. cerebri post., pars precommunicalis*

TA English: *Posterior cerebral artery, precommunicating part*

Corresponds to the P1 segment.

Extends from origin (division of the basilar artery) to the posterior communicating artery. Situated in the interpeduncular cistern.

Posterior cerebral artery, terminal part (cortical)

Vessels 3

TA Latin: *A. occipitalis med.*

TA English: *Medial occipital artery*

Corresponds to the P4 segment and thus to the final division of the posterior cerebral artery. It courses to the cerebral cortex where it divides into two vessels: lateral occipital artery and middle occipital artery. The latter then divides into the parieto-occipital branch and calcarine branch.

Posterior column

Pathways

TA Latin: *Funiculus post.*

TA English: *Posterior funiculus*

The cuneate fasciculus and gracile fasciculus together form the posterior column and are the main axes of epicritic sensibility:

– gracile fasciculus: it collects the epicritic fibers from the sacral, lumbar as well as lower

thoracic cord and terminates in the gracile nucleus.
– cuneate fasciculus: contains the fibers from the upper thoracic cord as well as from the cervical cord and terminates in the cuneate nucleus.

Posterior commissure

Telencephalon   3
TA Latin: *Commissura post.*
TA English: *Posterior commissure*
Here cross the fibers that are vital for controlling vertical eye movement and consensual light reaction of the pupils, including fibers from the superior colliculus, pretectal region as well as tegmentum of mesencephalon.

Posterior communicating artery

Vessels   3
TA Latin: *A. communicans post.*
TA English: *Posterior communicating artery*
Large connecting segment between the posterior cerebral artery and the internal carotid artery. Important component of the arterial circle of Willis.
Courses in the interpeduncular cistern.
Gives off short and small lateral branches in the direction of the mesencephalon (mesencephalic branch).

Posterior communicating artery, anteroinferior thalamic branch

Vessels   1
→ Posterior communicating artery, hypothalamic branch

Posterior communicating artery, hypothalamic branch

Vessels   1
TA Latin: *A. communicans post., R. hypothalamicus*
TA English: *Posterior communicating artery, hypothalamic branch*
In some cases, lateral branches emerge from the posterior communicating artery, supplying parts of the hypothalamus and thalamus. These branches are certainly not always encountered.

Posterior communicating vein

Vessels   3
Connects the right anterior vein and left anterior vein and is hence an important component of the venous circle of cerebrum, the so-called Trolard's hexagon.

Posterior ethmoidal artery

Vessels   1
TA Latin: *A. ethmoidalis post.*
TA English: *Posterior ethmoidal artery*
The artery passing through the ethmoid bone and arising from the ophthalmic artery and forming part of the orbitomeningeal anastomosis.

Posterior ethmoidal foramen

Meninges & Cisterns   1
TA Latin: *Foramen ethmoidale post.*
TA English: *Posterior ethmoidal foramen*
Point of passage for the similarly named vessels and nerves between the ethmoid bone and frontal bone and resting on the medial wall of the orbita.

Posterior external vertebral venous plexus

Vessels
TA Latin: *Plexus venosus vertebralis externus post.*
TA English: *Posterior external vertebral venous plexus*
Greatly anastomosed venous plexus outside the vertebral column. Via the basivertebral veins, they are connected with the internal vertebral venous plexus.

Posterior forceps

Telencephalon   2
TA Latin: *Forceps major*
TA English: *Major forceps*
The commissural fibers running in the splenium of the corpus callosum from the occipital lobe embark on a U-shaped course and, viewed in horizontal section, are shaped like forceps. They are called the posterior forceps.
The anterior forceps is formed from similar U-shaped fibers in the frontal lobe.

Posterior gray commissure
Medulla spinalis   2
TA Latin: *Commissura grisea post.*
TA English: *Posterior gray commissure*
In the gray commissure, the nuclear regions, more precisely the intermediate substance, of both halves of spinal cord meet each other.
Whereas the anterior gray commissure runs ventrally to the central canal, the posterior gray commissure passes dorsally to the spinal canal.

Posterior horn
Medulla spinalis   3
TA Latin: *Cornu post.*
TA English: *Posterior horn*
The majority of primary afferents entering through the posterior horn terminate in the posterior horn of the spinal cord. Three zones can be distinguished:
– marginal cells,
– substantia gelatinosa,
– nucleus proprius.

Posterior horn, marginal nucleus
→ Marginal cells

Posterior hypophyseal lobe
Diencephalon
TA Latin: *Neurohypophysis*
TA English: *Neurohypophysis*
→ Posterior lobe of the hypophysis

Posterior hypothalamic nucleus
Diencephalon   2
TA Latin: *Nucl. post. hypothalami*
TA English: *Posterior hypothalamic nucleus*
The nucleus contains a number of intrathalamic, mesencephalic and medullary afferents (e.g. locus coeruleus, habenula, lateral hypothalamic area, solitary nucleus, reticular formation). Efferents are virtually unknown (raphe nuclei, locus coeruleus, spinal cord). The entire posterior hypothalamus is involved in the generation of emotional stress and in regulation of heat production.

Posterior inferior cerebellar and vertebral artery, intracranial part
Vessels
TA Latin: *A. inf. post. cerebelli u. A. vertebralis, pars intracranialis*
TA English: *Posterior inferior cerebellar and vertebral artery, intracranial part*
→ Medullary branches

Posterior inferior cerebellar artery
Vessels   3
TA Latin: *A. inf. post. cerebelli*
TA English: *Posterior inferior cerebellar artery*
Greatest lateral branch of the intracranial part of the vertebral artery. Arises at the level of the accessory nerve and passes beneath the cerebellum, where it divides into several branches.
Supplies the underside of the cerebellar vermis as well as the under surface of the cerebellar hemispheres.

Posterior inferior cerebellar artery, cerebellar tonsillar branch
Vessels   1
TA Latin: *A. inf. post. cerebelli, R. Tonsillae cerebelli*
TA English: *Posterior inferior cerebellar artery, cerebellar tonsillar branch*
Lateral branch of the medial branch of the posterior inferior cerebellar artery.
Supplies the tonsil of cerebrum.

Posterior inferior cerebellar artery, choroid branch of fourth ventricle
Vessels   2
TA Latin: *A. inf. post. cerebelli, R. Choroideus ventriculi quarti*
TA English: *Posterior inferior cerebellar artery, choroidal branch to fourth ventricle*
Together with the ventricular branches, the inferior artery of vermis, and various other branches, it travels in the direction of the underside of the cerebellum where it forms the terminal segments of the posterior inferior cerebellar artery.
As suggested by the name, the lateral branch supplies the choroid plexus of the fourth ventricle.

Posterior inferior cerebellar artery, lateral branch

Vessels   2

Large terminal segment of the posterior inferior cerebellar artery which courses laterally around the cerebellum and supplies the cortical regions of the cerebellar hemispheres.

Posterior inferior cerebellar artery, lateral medullary branches

Vessels   1

→ Posterior inferior cerebellar artery, medial medullary branches

Posterior inferior cerebellar artery, medial branch

Vessels   2

Large terminal segment of the posterior inferior cerebellar artery which courses medially around the cerebellum and supplies the cortical regions of the cerebellar hemispheres, including the tonsil via the cerebellar tonsillar branch.

Posterior inferior cerebellar artery, medial medullary branches

Vessels   1

On its way to the cerebellum, the posterior inferior cerebellar artery gives off a variable number of side branches in the direction of the myelencephalon. These branches are called the medullary branches, with a distinction being made between the medial, posterior and lateral branches.

Posterior inferior cerebellar artery, posterior medullary branches

Vessels

→ Posterior inferior cerebellar artery, medial medullary branches

Posterior intercavernous sinus

Vessels   3

TA Latin: *Sinus intercavernosus post.*
TA English: *Posterior intercavernous sinus*
The right and left cavernous sinuses are connected via the short anterior intercavernous sinuses and posterior intercavernous sinus. The sinus ring thus formed is called the circular si-

nus and it surrounds the hypophysis. It receives its venous incoming blood from the latter, as well as from the sphenoid sinus and diaphragma sellae.

Posterior intercostal artery

Vessels   1

TA Latin: *A. intercostalis post.*
TA English: *Posterior intercostal artery*
The posterior intercostal arteries arise from the thoracic aorta and supply the back muscles, spinal cord, spinal meninges, intercostal spaces, skin and mammary glands.
They give off branches both dorsally (dorsal roots) and to the spinal ganglia (spinal branches).

Posterior intercostal vein

Vessels   3

TA Latin: *V. intercostalis post.*
TA English: *Posterior intercostal vein*
Veins coursing in the costal sulcus conducting blood from the back musculature, vertebral canal and vertebral column to the azygos vein and hemiazygos vein.

Posterior intermediate sulcus

Myelencephalon   2

TA Latin: *Sulcus intermedius post.*
TA English: *Posterior intermediate sulcus*
Typical sulcus on the dorsal side of the spinal cord. Separates cuneate fasciculus and gracile fasciculus from each other.

Posterior internal frontal artery

Vessels

TA Latin: *A. callosomarginalis, R. Front. posteromed.*
TA English: *Callosomarginal artery, posteromedial frontal branch*
→ Callosomarginal artery, posteromedial frontal branch

Posterior internal vertebral venous plexus

Vessels

TA Latin: *Plexus venosus vertebralis internus post.*
TA English: *Posterior internal vertebral venous plexus*

Greatly anastomosed anterior internal vertebral venous plexus and posterior internal vertebral venous plexus are venous plexuses in the vertebral canal. They collect blood from the vertebral canal and transport it via the venous plexuses of the intervertebral canals to the veins outside the vertebral canal.

Via the basivertebral veins they are connected with the posterior external vertebral venous plexus.

Posterior lateral thalamic nucleus

Diencephalon   2
TA Latin: *Nucl. lat. post. thalami*
TA English: *Lateral posterior nucleus of thalamus*
Like the anterior part of the thalamic pulvinar, this thalamic nucleus of the lateral nuclear group features reciprocal connections with area 5 and area 7 (secondary somatosensory cortex). Its tasks are in the domain of pain conduction, somatosensory control, motor control and gustation.

Posterior limb of internal capsule

Telencephalon   3
TA Latin: *Capsula interna, crus post.*
TA English: *Posterior limb of internal capsule*
The internal capsule features the following pathways:
posterior limb of internal capsule:
– pyramidal tract,
– superior thalamic peduncle,
– posterior thalamic peduncle,
– parietopontine tract,
– corticotegmental fibers,
Anterior limb of internal capsule:
– frontopontine tract,
– anterior thalamic peduncle.

Posterior lobe of the hypophysis

Diencephalon   3
TA Latin: *Neurohypophysis*
TA English: *Neurohypophysis*
The posterior lobe of the hypophysis is also called the neurohypophysis since it is composed of hypothalamic nervous tissue. Its proximal segment is formed by the tuber cinerum and infundibulum, and its distal segment is the posterior lobe of the hypophysis. Via the infundi-

bular nucleus, axons of the paraventricular nucleus and of the supraoptic nucleus pass to the blood vessels in the posterior lobe, where they release the hormones ADH and oxytocin.

Posterior longitudinal ligament

    1
TA Latin: *Ligamentum longitudinale post.*
TA English: *Posterior longitudinal ligament*
The posterior longitudinal ligament of the vertebral column runs in the vertebral arches and is firmly connected with the intervertebral discs.

Posterior median septum

Medulla spinalis   2
TA Latin: *Septum medianum post.*
TA English: *Posterior median septum*
→ Posterior median sulcus

Posterior median sulcus

Medulla spinalis   3
TA Latin: *Sulcus medianus post.*
TA English: *Posterior median sulcus*
A longitudinal sulcus, called the posterior median sulcus, stretches on the dorsal side in the middle of the spinal cord. This continues to a deeper level where it runs as the posterior median septum almost as far as the central canal.

Posterior meningeal artery

Vessels   2
TA Latin: *A. meningea post.*
TA English: *Posterior meningeal artery*
Arises from the ascending pharyngeal artery and passes via the jugular foramen into the posterior cranial fossa, where it supplies the dura.

Posterior nucleus of hypothalamus

Diencephalon   1
TA Latin: *Nucl. post. hypothalami*
TA English: *Posterior nucleus of hypothalamus*
→ Posterior hypothalamic nucleus

Posterior parietal artery

Vessels   2
TA Latin: *A. parietalis post.*
TA English: *Posterior parietal artery*
Arises from the middle cerebral artery, parietal trunk. The middle cerebral artery for its part emerges from the internal carotid artery.

Supplies the inferior parietal lobule, parts of the parietal lobe as well as the upper section of the postcentral gyrus.

Posterior parolfactory sulcus

Telencephalon   2
TA Latin: *Sulcus parolfactorius post.*
TA English: *Posterior parolfactory sulcus*
The subcallosal area is enclosed by the anterior parolfactory sulcus and posterior parolfactory sulcus.

Posterior perforated substance

Mesencephalon   2
TA Latin: *Subst. perforata post.*
TA English: *Posterior perforated substance*
The posterior perforated substance lies deep in the interpeduncular fossa, where the oculomotor nerve (III) emerges.

Posterior pretectal nucleus

Mesencephalon   1
TA Latin: *Nucl. pretectalis post.*
TA English: *Posterior pretectal nucleus*
→ Pretectal area

Posterior radicular artery

Vessels   2
TA Latin: *A. radicularis post.*
TA English: *Posterior radicular artery*
These are short arterial branches from the vertebral artery, which in the cervical region supply the spinal ganglia and the ventral and dorsal roots of the spinal nerve.

Posterior radicular vein

Vessels   3
Veins penetrating the dorsal root into the spinal canal and thus creating an anastomosis between the hemiazygos vein and the posterior spinal vein which courses along the spinal cord.

Posterior spinal artery

Vessels   3
TA Latin: *A. spinalis post.*
TA English: *Posterior spinal artery*
The unpaired anterior spinal artery and the paired posterior spinal artery arise from the vertebral artery and supply the spinal cord and spinal meninges.

Posterior spinal vein

Vessels
TA Latin: *V. spinalis post.*
TA English: *Posterior spinal vein*
→ Spinal veins (anterior, lateral, posterior)

Posterior spinocerebellar tract

Cerebellum   3
TA Latin: *Tractus spinocerebellaris post.*
TA English: *Posterior spinocerebellar tract*
The posterior spinocerebellar tract carries primary afferents from the spinal cord to the cerebellum.
It has its origin in Clarke´s column in the thoracic cord and conducts proprio- and exteroceptive impulses (skin receptors, muscle spindles, tendon spindles) from the posterior limbs to the cerebellum.

Posterior superior fissure

Cerebellum   2
TA Latin: *Fissura post. sup.*
TA English: *Posterior superior fissure*
The posterior superior fissure has its origin in the folium vermis and separates the simple lobule and superior semilunar lobule.

Posterior tegmental venulae

Vessels   1
These small veins drain the area around the vestibular nucleus and carry the venous blood into the lateral mesencephalic vein

Posterior temporal artery

Vessels   2
TA Latin: *R. temporalis post. a. cerebri med.*
TA English: *Posterior temporal branch of a. cerebri med.*
Arises from the middle cerebral artery, posterior trunk. The middle cerebral artery emerges from the internal carotid artery.
Supplies the lower portion of the temporal lobe, transverse temporal gyri as well as the sensory speech center (Wernicke's area).

Posterior thalamic branch

Vessels 1

The posterior thalamic branch arises from the postcommunical part of the posterior cerebral artery.

This branch supplies the posterior thalamic complex such as the geniculate bodies but also the contiguous regions such as internal capsule.

Posterior thalamic nuclei

Diencephalon 3

TA Latin: *Nuclei post. thalami*

TA English: *Posterior nuclear complex of thalamus*

The posterior part of the thalamus contains the thalamic pulvinar and the nuclei of the metathalamus. Its tasks are in the visual and acoustic signal processing as well as in visuomotoric integration.

Posterior thalamic nucleus

Diencephalon

TA Latin: *Nucl. post. thalami*

TA English: *Posterior nucleus of thalamus*

→ Posterior thalamic nuclei

Posterior trunk of the middle cerebral artery

Vessels

→ Middle cerebral artery, posterior trunk

Posteroinferior diencephalic branches

Vessels

TA Latin: *Aa. centrales posteromed.*

TA English: *Posteromedial central arteries*

→ Posteromedial central arteries

Posteroinferior diencephalic branches; deep interpeduncular branches

Vessels

TA Latin: *Aa. centrales posteromed.*

TA English: *Posteromedial central arteries*

→ Posteromedial central arteries

Posterolateral column

Medulla spinalis 2

TA Latin: *Tractus posterolat. (Lissauer)*

TA English: *Posterolateral tract (Lissauer)*

The white matter between the ventral root and dorsal root gives rise to the lateral column, containing:

1) anterolateral column with
– anterolateral fasciculus
– parts of the anterior spinocerebellar tract.
2) posterolateral column with
– posterior spinocerebellar tract
– parts of the anterior spinocerebellar tract
– lateral pyramidal tract.

Posterolateral fissure

Cerebellum 2

TA Latin: *Fissura posterolat.*

TA English: *Posterolateral fissure*

The posterolateral fissure separates the nodulus and uvula vermis.

Posterolateral sulcus

Medulla spinalis 2

TA Latin: *Sulcus lat., R. post.*

TA English: *Lateral sulcus, posterior ramus*

Dorsolateral sulcus of the spinal cord. It separates the cuneate fasciculus and lateral funiculus from each other.

Posteromedial central arteries

Vessels 1

TA Latin: *Aa. centrales posteromed.*

TA English: *Posteromedial central arteries*

Small lateral branches arising from the precommunical part of the posterior cerebral artery and supplying the surrounding brain tissue, primarily the subthalamus.

Posteromedial central arteries, subthalamic area

Vessels

TA Latin: *Aa. centrales posteromed.; area subthalamica*

TA English: *Posteromedial central arteries, subthalamic area*

→ Posteromedial central arteries

Precentral artery

Vessels

TA Latin: *A. sulci precentralis*

TA English: *Artery of precentral sulcus*

→ Artery of precentral sulcus

Precentral fissure

Cerebellum   2
TA Latin: *Fissura precentralis*
TA English: *Precentral fissure*
Cerebellar fissure situated rostral to the central lobule.

Precentral gyrus (area 4)

Telencephalon   3
TA Latin: *Gyrus precentralis (Area 4)*
TA English: *Precentral gyrus (area 4)*
= Motor cortex= area 4 = primary somatomotor cortex. The precentral gyrus lies in the frontal lobe, directly on the central sulcus. The pyramid cells are encountered here, providing motor control for the contralateral skeletal muscles. All voluntary movements are conducted via this gyrus. It has strict somatotopic arrangement and passes into the longitudinal fissure of cerebrum.
Damage to the precentral gyrus alone results in flaccid paralysis of the contralateral skeletal musculature. If premotor areas are concurrently affected, spastic paralysis can ensue, as inhibitory influence on the centers of brainstem and thalamus are lacking, hence increased muscle tone of the extrapyramidal system is predominant.

Precentral sulcus

Telencephalon   3
TA Latin: *Sulcus precentralis*
TA English: *Precentral sulcus*
The precentral sulcus stretches in the frontal lobe almost parallel to the central sulcus and separates the motor cortex from the premotor cortex.

Precentral vein of cerebellum

Vessels   1
TA Latin: *V. precentralis cerebelli*
TA English: *Precentral cerebellar vein*
The precentral cerebellar veins course along the cerebellar lobule and conveys its blood in the direction of the straight sinus, while various intermediate vessels may be encountered.

Precommissural components of the thalamic medullary stria

Diencephalon   1

Above the anterior commissure, the fiber bundles from the hypothalamus join the thalamic medullary stria, with some ascending before (precommissural) and some after (postcommissural) the anterior commissure.
These components course together with the precommissural stria terminalis and the precommissural fornix.

Precommissural fornix

Diencephalon   1
TA Latin: *Fibrae precommissurales columnae fornicis*
TA English: *Precommissural fibres of column of fornix*
Part of the fornix that branches off above the anterior commissure and enters the septal verum.

Precommissural septum

→ Septum verum

Preculminate fissure

Cerebellum   2
TA Latin: *Fissura preculminalis*
TA English: *Preculminate fissure*
The preculminate fissure separates the quadrigeminal lobule and the ala lobuli centralis

Precuneal artery

Vessels   2
Together with the paracentral artery, it arises from the internal parietal artery. Passes to the precuneus, which it also supplies.
Together with the posterior branch, the internal parietal artery emerges from the pericallosal artery.
There are also branches emerging from the pericallosal artery, posterior branch, which have the same function.

Precuneal branches

Vessels
TA Latin: *Rr. precuneales*
TA English: *Precuneal branches*
→ Precuneal artery

Precuneal branches; internal parietal artery (inferior and superior)

Vessels

TA Latin: *A. pericallosa, Rr. precuneales*
TA English: *Pericallosal artery, precuneal branches*
→ **Precuneal artery**

Precuneus

Telencephalon 2
TA Latin: *Precuneus*
TA English: *Precuneus*
Part of the parietal lobe visible in a median section. Has a virtually square shape (hence also called quadrate lobe). The precuneus appears to be implicated in complex, sensory evaluation processes, language processing as well as spatial and temporal orientation.

Prefrontal artery (candelabra artery)

Vessels 1
TA Latin: *A. prefrontalis*
TA English: *Prefrontal artery*
Arises from the middle cerebral artery, frontal trunk. The middle cerebral artery for its part emerges from the internal carotid artery.
Supplies the inferior frontal gyrus, triangular part.

Prefrontal cortex

Telencephalon 2
Integrated cortical area. Afferents from the medial thalamic nucleus and from the perirhinal cortex. Concomitantly, long association systems from the mulitmodal regions in the occipital, temporal and parietal lobes terminate here, projecting inter alia, to the amygdaloid body and the postcentral association cortex.
Has very close links to limbic and paralimbic association regions. The orbitofrontal cortex is also a component of this cortical portion.

Prefrontal veins

Vessels 2
TA Latin: *Vv. prefrontales*
TA English: *Prefrontal veins*
Superficially coursing veins on the anterior frontal lobe.

Pregeniculate nucleus

Diencephalon 2
TA Latin: *Nucl. ventr. corporis geniculati lat.*
TA English: *Ventral lateral geniculate nucleus*

This nucleus lies dorsally to the lateral geniculate body and receives afferents directly from the retina. Further afferents come from the suprachiasmatic nucleus as well as the contralateral pregeniculate nucleus. Efferents go to the pineal body.

Prelimbic area (area 32)

Telencephalon 1
Frontal cortex area which is contiguous with the limbic lobe.

Prelimbic cortex (area 32)

Telencephalon 1
Areas of the frontal lobe which are contiguous with the limbic system.

Premamillary artery

Vessels
TA Latin: *A. communicans post., R. hypothalamicus*
TA English: *Posterior communicating artery, hypothalamic branch*
→ **Posterior communicating artery, hypothalamic branch**

Premotor area

Telencephalon
→ **Frontal cortex (areas 6+8)**

Premotor cortex (area 6)

Telencephalon 2
The premotor cortex (area 6) appears to play a role in regulating grasping actions. In the caudal segments fibers arise from the pyramidal tract. Lesions frequently result in impaired grasping actions. Strength regulation of the grasping hand is likewise impaired.

Premotor cortex (area 8)

Telencephalon 2
Frontal eye field. Plays an important role in voluntary control of eye movement.
Lesions in this area cause loss of voluntary control of eye movement.

Preoccipital notch

Telencephalon 2
TA Latin: *Incisura preoccipitalis*

TA English: *Preoccipital notch*
Short but pronounced notch, separating the temporal lobe form the occipital lobe

Preoptic area

Diencephalon　　　　　　　　　　3
TA Latin: *Area preoptica*
TA English: *Preoptic area*
Situated at the lateral wall of the third ventricle, close to the optic recess. The region comprises three nuclear areas: periventricular nucleus, medial preoptic nucleus and lateral preoptic nucleus.
The area is also in the direct vicinity of the organum vasculosum of the lamina terminalis and plays a role in thermoregulation, hypovolemic thirst, male sexual behavior, brood care, gonadotropin secretion and locomotion.

Preoptic nucleus

Diencephalon　　　　　　　　　　1
TA Latin: *Nucl. preopticus*
TA English: *Preoptic nucleus*
The lateral preoptic nucleus is involved in locomotion.
The medial preoptic nucleus, conversely, is involved in thermoregulation, hypovolemic thirst, male sexual behavior, nursing care, modulation of gonadotropin secretion.

Preoptic region

TA Latin: *Area preoptica*
TA English: *Preoptic area*
→ Preoptic area

Prepiriform cortex

Telencephalon　　　　　　　　　　1
This cortical area belongs to the basal olfactory area. The medial part lies on the base of the frontal lobe above the lateral olfactory stria, the cerebellar hemisphere, lateral part is located in the temporal lobe. Caudally, the superficial cortical amygdaloid nucleus is encountered instead of the prepiriform cortex.
Efferents to the medial thalamic nucleus, substantia innominata and olfactory bulb.

Prepositus hypoglossal nucleus

Myelencephalon　　　　　　　　　　2

TA Latin: *Nucl. prepositus hypoglossi*
TA English: *Prepositus hypoglossal nucleus*
The rod-shaped nucleus rostral to the hypoglossal nerve in the oculomotor control center which plays an important role in tracking moving objects with the eyes, coordination of fast eye movements and fixation on objects. The nucleus has afferents from the cerebellum, vestibular nuclei as well as the interstitial nucleus (Cajal). Efferents to the nuclei of the eye muscles.

Prepyramidal fissure

Cerebellum　　　　　　　　　　2
TA Latin: *Fissura prebiventralis*
TA English: *Prebiventral fissure*
Cerebellar fissure before the pyramid of vermis.

Presubiculum

Telencephalon　　　　　　　　　　1
TA Latin: *Presubiculum*
TA English: *Presubiculum*
Parasubiculum and presubiculum belong to the hippocampal region and evidence a (inhomogeneous) 4-layered structure.
Afferents come from the olfactory system, dorsal thalamus, as well as the contralateral presubiculum.
Efferents go to the entorhinal area, hippocampus, mammillary body and tegmentum of mesencephalon.

Pretectal area

Mesencephalon　　　　　　　　　　2
TA Latin: *Area pretectalis*
TA English: *Pretectal area*
Situated immediately behind the superior colliculus, this nucleus plays a vital role in pupillary reflex and adaptation.
Afferents come from the retina and occipital cortical fields. Efferents go to the ipsi- and contralateral accessory oculomotor nucleus and superior colliculus.
The pretectal area includes the following nuclei: pretectal olivar nucleus, medial, anterior and posterior pretectal nuclei and optic tract nucleus.

Pretectal olivary nucleus

Mesencephalon　　　　　　　　　　1
TA Latin: *Nucl. pretectalis olivaris*

TA English: *Olivary pretectal nucleus*
→ Pretectal area

Pretectal region
TA Latin: *Area pretectalis*
TA English: *Pretectal area*
→ Pretectal area

Pretectum
TA Latin: *Pretectum*
TA English: *Pretectum*
→ Pretectal area

Primary acoustic cortex
Telencephalon 3
Cerebral cortex areas in which the auditory tract terminates and which are involved in the first cortical processing steps for auditory signals. These include especially Brodmann areas 41 and 42 on the temporal plane.

Primary afferents
General CNS 1
Primary afferents are tracts ascending without interneurons.
One example is the posterior spinocerebellar tract which conducts impulses without interneurons from Clarke's nucleus of the thoracic cord to the cerebellar hemispheres.

Primary fissure
Cerebellum 3
TA Latin: *Fissura prima*
TA English: *Primary fissure*
The primary fissure separates the anterior cerebellar lobe from the posterior lobe.

Primary motor cortex
Telencephalon
TA Latin: *Gyrus precentralis*
TA English: *Precentral gyrus*
→ Precentral gyrus (area 4)

Primary somatosensory cortex
Telencephalon
= SI
= area 3 + 1 + 2
The postcentral gyrus lies in the parietal lobe directly on the central sulcus. Observing strict somatotopic arrangement, the somatosensory tracts of the contralateral body half terminate here.
Conscious localization and differentiation of quality and intensity of a tactile stimulus are effected in cooperation with the postcentral gyrus.

Primary visual cortex
Telencephalon
→ Area 17 (striate cortex)

Principal mammillary fasciculus
Diencephalon 3
Exit site of the mammillary efferents directly at the mammillary body. The fasciculus then divides into the mammillothalamic fasciculus, Vicq d'Azyr bundle and the mammillotegmental tract.

Principal olivary nucleus
TA Latin: *Nucl. olivaris principalis*
TA English: *Principal olivary nucleus*
→ Inferior olive

Principal sensory nucleus of the trigeminal nerve
Mesencephalon 2
TA Latin: *Nucl. principalis n. trigemini*
TA English: *Principal sensory nucleus of trigeminal nerve*
This nucleus lies dorsal to the lateral recess of the fourth ventricle in the tegmentum of pons. Afferents are the axons of cells of the trigeminal ganglion with impulses from the forehead and face, masticatory musculature, eye muscles and mucosa.
Efferents go to the thalamus: via the ventral tegmental fasciculus to the contralateral side and via the dorsal trigeminothalamic tract to the ipsilateral ventral posteromedial nucleus.

Probst's intersection
→ Decussation of the lateral lemnisci

Projection tracts
General CNS
Commissures are fibers which exchange information between the hemispheres. Association pathways are fiber bundles within a hemisphere,

while fibers between cerebral cortex and sub-cortical centers are called projection pathways or tracts.

Proprioceptive stimuli

As regards the somatosensory control, there are two groups of sensibility:
- the protopathic sensibility includes crude touch and pressure perceptions, pain and temperature.
- the epicritic sensibility comprises extero-ceptive stimuli (exact, tactile stimuli of the mechanoreceptors of the skin) and proprio-ceptive stimuli (position of the body in space, mediated by joint and muscle receptors).

Prosencephalon

General CNS   3
TA Latin: *Prosencephalon*
TA English: *Prosencephalon*
Composed of telencephalon and diencephalon.

Protopathic sensibility   3

As regards the somatosensory control, there are two groups of sensibility:
- the protopathic sensibility includes crude touch and pressure perceptions, pain and temperature.
- the epicritic sensibility comprises extero-ceptive stimuli (exact, tactile stimuli of the mechanoreceptors of the skin) and proprio-ceptive stimuli (position of the body in space, mediated by joint and muscle receptors).

Pterygoid plexus

Vessels   1
TA Latin: *Plexus pterygoideus*
TA English: *Pterygoid plexus*
Venous plexus resting on the lateral and medial pterygoid muscles.

Pterygoid venous plexus

Vessels
TA Latin: *Plexus pterygoideus*
TA English: *Pterygoid plexus*
→ Pterygoid plexus

Pulvinar nuclei

Diencephalon   3
TA Latin: *Nuclei pulvinares*
TA English: *Pulvinar nuclei*
The thalamic pulvinar belongs to the lateral cell group of the thalamus and comprises four parts, which are also designated as nuclei:
- thalamic pulvinar, anterior part,
- thalamic pulvinar, lateral part
- thalamic pulvinar, medial part,
- thalamic pulvinar, inferior part

The functions of this part of the thalamus are not clearly understood, but their role appears to be in the somatosensory/somatomotor domain.

Pupil

Eye   1
TA Latin: *Pupilla*
TA English: *Pupil*
Part of the eye. Important for adaptation.

Purkinje cells

Cerebellum   3
TA English: *Purkinje cells*
Large, pear-shaped cells in the cerebellar cortex, featuring 2-3 horizontally extending dendrites and one axon, generally projecting to a cerebel-lar nucleus.

Putamen

Telencephalon   3
TA Latin: *Putamen*
TA English: *Putamen*
The caudate nucleus and putamen together form the corpus striatum. Both are derived ontogenetically from the same anlagen, but are separated by incoming fibers from the internal capsule.
The corpus striatum is an important inhibitory component of motor movement programs and has manifold connections with the globus palli-dus, substantia nigra and the motor cortex.

Pyramid of medulla oblongata

Myelencephalon   2
TA Latin: *Pyramis medullae oblongatae*
TA English: *Pyramid of medulla oblongata*
On the dorsal side of the brainstem, a well-delineated myelinated tuberculum can be dis-

cerned at the transition between spinal cord and pons. Here the pyramidal tract rises to the surface before it soon divides into two strands at the pyramidal decussation.

Pyramid of vermis

Cerebellum 2

TA Latin: *Pyramis vermis (VIII)*
TA English: *Pyramid of vermis (VIII)*
A segment of the vermis cerebelli between the biventer lobules.
Like the entire vermis cerebelli, the pyramid of vermis receives its afferents primarily from the spinal cord. It is also part of the spinocerebellum = palaeocerebellum.

Pyramidal decussation

Myelencephalon 3

TA Latin: *Decussatio pyramidum*
TA English: *Decussation of pyramids*
In the pyramidal decussation, 70–90% of the pyramidal tract decussate to the contralateral side. The decussation lies directly below the pyramid (of the myelencephalon), in the middle of the myelencephalon. The decussation joins the lateral pyramidal tract.

Pyramidal tract

Nerves/Pathways 3

TA Latin: *Tractus pyramidalis*
TA English: *Pyramidal tract*
The largest descending motor tract is formed by the axons of the pyramidal cells of the motor cortex and is thus called the pyramidal tract. It courses nonstop from the cortex to the corresponding segments in the spinal cord, explaining why it is also called the corticospinal tract. At the upper margin of the myelencephalon, this tract rises to the surface as the pyramid (of myelencephalon).
70–90% of the fibers then cross in the pyramidal decussation to the contralateral side, and continue to run in the spinal cord as the lateral pyramidal tract. The remaining fibers descend in the medial pyramidal tract. The fibers project directly or indirectly to the alpha motoneurons, especially of the distal extremities (hand/forearm), hence the pyramidal tract plays a vital role in fine motor control.

Pyramidal tract of cerebral peduncle

Pathways 2

TA Latin: *Tractus pyramidalis*
TA English: *Pyramidal tract*
The pyramidal tract is an important component of the cerebral peduncles.

Q

Quadrangular lobule

Cerebellum   2

TA Latin: *Lobulus quadrangularis*

TA English: *Quadrangular lobule*

The ventral hemisphere section belongs to the posterior lobe.

Apart from the areas in proximity to the vermis (cerebellar hemisphere, intermediate part), the hemispheres belong to the phylogenetically young neocerebellum and receive their afferents via the mossy fibers of the pontocerebellar tract from the pontine nuclei. All hemisphere segments are hence also assigned to the pontocerebellum

Quadrate lobe

Telencephalon

TA Latin: *Lobus quadratus*

TA English: *Quadrate lobe*

Part of the parietal lobe visible in a median section. Has a virtually square shape (hence also called quadrate lobe). The precuneus appears to be implicated in complex, sensory evaluation processes, language processing as well as spatial and temporal orientation.

Quadrigeminal artery

Vessels   2

TA Latin: *A. collicularis*

TA English: *Collicular artery*

From the precommunical part of the posterior cerebral artery arises a series of larger vessels including the quadrigeminal artery, which passes through the cisterna ambiens to the quadrigeminal plate, supplying there with its branches the tegmentum of mesencephalon, geniculate bodies and – as suggested by the name – the quadrigeminal lamina.

Quadrigeminal cistern

Meninges & Cisterns

TA Latin: *Cisterna quadrigeminalis*

TA English: *Quadrigeminal cistern*

→ Cistern of tectal lamina

Quadrigeminal plate

Mesencephalon   2

TA Latin: *Lamina tecti*

TA English: *Tectal plate*

Also called tectum or quadrigeminal plate. Composed of two pairs of hills:

superior colliculus: the two upper hills belong to the visual system (control of eye movements).

inferior colliculus: the two lower hills belong to the auditory system and are an integral part of information exchange from inner ear to auditory cortex.

R

Radiation of the corpus callosum
Telencephalon   2
TA Latin: *Radiatio corporis callosi*
TA English: *Radiation of corpus callosum*
Radiation of the corpus callosum is the term used for the ray-shaped fiber course from the corpus callosum to the cortical regions of the telencephalon.

Radicular branches
Vessels
TA Latin: *Rr. radiculares*
TA English: *Radicular branches*
→ Anterior radicular artery

Raphe magnus nucleus (B3)
TA Latin: *Nucl. raphes magnus (B3)*
TA English: *Magnus raphe nucleus (B3)*
→ Raphe nuclei

Raphe nuclei
Mesencephalon   3
TA Latin: *Nuclei raphes*
TA English: *Raphe nuclei*
The raphe nuclei form the median zone of the reticular formation, belong to the mono-aminergic cell groups and comprise 6 serotoninergic nuclear regions: raphe pallidus nucleus (B1), raphe obscurus nucleus (B2), raphe magnus nucleus (B3), raphe pontine nucleus (B5), dorsal raphe nucleus (B7) and central superior raphe nucleus. They constitute the largest neuronal network known.
Afferents: noradrenergic systems, limbic system and cerebellum.
Efferents: all regions of the brain.

Raphe obscurus nucleus (B2)
TA Latin: *Nucl. raphes obscurus (B2)*

TA English: *Obscurus raphe nucleus (B2)*
→ Raphe nuclei

Raphe pallidus nucleus (B1)
TA Latin: *Nucl. raphes pallidus (B1)*
TA English: *Pallidal raphe nucleus (B1)*
→ Raphe nuclei

Raphe pontine nucleus (B5)
TA Latin: *Nucl. raphes pontis (B5)*
TA English: *Pontine raphe nucleus (B5)*
→ Raphe nuclei

Raphespinal tract
Pathways   2
TA Latin: *Tractus raphespinalis*
TA English: *Raphespinal tract*
Projections of the magnocellular raphe nuclei (median zone of the reticular formation) to the gray matter of the spinal cord.

Recurrent artery
Vessels   3
TA Latin: *A. striata med. distalis (Heubneri)*
TA English: *Distal medial striate artery (Heubner)*
→ Long central artery (Heubner´s)

Red nucleus
Mesencephalon   3
TA Latin: *Nucl. ruber*
TA English: *Red nucleus*
This nucleus, named for its high content of red-colored iron, lies in the tegmentum of mesencephalon and is a component of the motor system. It receives afferents from the contralateral central dentate nucleus and emboliform nucleus (cerebellum), as well as from the ipsilateral globus pallidus and cerebral cortex. Efferents: spinal cord, reticular formation, olive. The red nucleus plays an important role in fine regulation of effector motor control and of muscle tone.
Lesions in the red nucleus detract from the basic tone of body musculature and produce intention tremor and chorea-like movements.

Red nucleus, magnocellular part
Mesencephalon   2

TA Latin: *Nucl. ruber, pars magnocellularis*
TA English: *Red nucleus, magnocellular part*
There are two cell populations in the red nucleus:
- red nucleus, magnocellular part: only poorly developed in humans. Fibers of the emboliform nucleus terminate here.
- red nucleus, parvocellular part: afferents from the dentate nucleus and cerebral cortex. Efferents to the olive.

Red nucleus, parvocellular part

Mesencephalon   2
TA Latin: *Nucl. ruber, pars parvocellularis*
TA English: *Red nucleus, parvocellular part*
→ Red nucleus, magnocellular part

Reflex eye movements, interruption

Damage to the superior colliculi results in interruption of reflex eye movements, but not in impairment of cognitive perceptivity (e.g. image recognition).

Release and release-inhibiting factors

→ Infundibular nucleus

Release of milk during lactation

→ Supraoptic nucleus

Restiform body

Cerebellum   2
TA Latin: *Corpus restiforme*
TA English: *Restiform body*
→ Inferior cerebellar peduncle

Reticular formation

General CNS   3
TA Latin: *Formatio reticularis*
TA English: *Reticular formation*
The reticular formation is an enormous network of cells filling the central portion of the brainstem, joining the intermediate substance of the spinal cord caudally, and rostrally the intralaminar thalamic nuclei (e.g. zona incerta). Three longitudinal zones are distinguished:
(1) median zone = raphe nuclei,
(2) medial, gigantocellular zone = medial reticular formation and

(3) a lateral (parvocellular) zone = lateral reticular formation. The RF is involved in gastrointestinal, cardiovascular and respiratory control, reflexes of the cranial nerves and pain suppression.

Reticular process

Medulla spinalis   1
Small processes of the intermediate substance. Since the intermediate substance can be considered to be part of the reticular formation, this process is called the reticular process.

Reticulo-parvocellular area

Myelencephalon   2
Small-celled portion of the reticular formation. The sum of these small-celled areas is called the lateral reticular formation and is involved in a series of vital autonomic processes.

Reticulocerebellar tract

Cerebellum   3
The reticulocerebellar tract has its origin mainly in the nucleus of the lateral funiculus (lateral funicle nucleus) and terminates chiefly in the vermis cerebelli, less so in the hemispheres.
The nucleus of the lateral funiculus in turn receives information via the spinoreticular afferents on tactile stimuli, whereas the receptive fields of the corresponding neurons can be distributed across large areas (e.g. both extremities).

Reticulospinal fibers

Pathways
TA Latin: *Fibrae reticulospinales*
TA English: *Reticulospinal fibres*
→ Reticulospinal tract

Reticulospinal tract

Pathways   3
TA Latin: *Tractus reticulospinalis ant.*
TA English: *Anterior reticulospinal tract*
The medial reticulospinal tract begins in the caudal pontine reticular nucleus and in the caudal portion of the oral pontine reticular nucleus. It descends in the medial longitudinal fasciculus in the spinal cord. Its fibers terminate mostly in lamina VII and VIII of the spinal gray matter;

but they also run in lamina IX in which the motoneurons for the trunk musculature lie.

Retina

Eye   3
TA Latin: *Retina*
TA English: *Retina*
Retina has a typical, multi-layer structure. The axons of the ganglion cells exit from the retina in the blind spot and course as the optic nerve in the direction of the brain.

Retinal ganglion cells

Eye
→ Retina

Retinohypothalamic fibers

Diencephalon   3
Individual fibers branch off from the optic tract and project directly to the supraoptic nucleus, which plays an important role in tuning certain circadian rhythms.

Retinohypothalamic pathway

Diencephalon   2
TA Latin: *Tractus retinohypothalamicus*
TA English: *Retinohypothalamic tract*
Fibres from the retina to the suprachiasmatic nucleus of the hypothalamus. Tasks in the regulation of circadiane rhythms.

Retroambiguus nucleus

Myelencephalon   1
TA Latin: *Nucl. retroambiguus*
TA English: *Retro-ambiguus nucleus*
Nuclear region of the myelencephalon continuing to the upper cervical cord and integrated in cardiorespiratory functions.

Retrocollis

Damage to the corpus striatum results in the typically manifest symptoms of chorea, due to disinhibition of the globus pallidus and substantia nigra. Chorea is characterized at an advanced stage by hyperkinesia especially of the distal extremities' musculature and of the face. Dystonic syndrome (e.g. retrocollis, spastic torticollis) or athetosis are also encountered.

Retrolenticular part of internal capsule

Telencephalon   2
TA Latin: *Capsula interna, pars retrolentiformis*
TA English: *Retrolentiform limb of internal capsule*
The portion of the internal capsule situated on the upper part of the tail of caudate nucleus, retrolenticular part contains, inter alia, the optic radiation as well as the occipitopontine tract.
The portion situated in the lower part, sublentiform part, conversely contains parts of the auditory radiation as well as the temporopontine tract.

Retromandibular vein

Vessels   2
TA Latin: *V. retromandibularis*
TA English: *Retromandibular vein*
Carries venous blood from the maxillary veins and the superficial temporal veins as well as other veins into the internal jugular vein. Its catchment area is thus the temples, auricle, parotid gland, maxilla and mandible as well as dura mater.

Retrosplenial cistern

Meninges & Cisterns
TA Latin: *Cisterna pericallosa*
TA English: *Pericallosal cistern*
→ Pericallosal cistern

Retrosplenial cortex (area 29+30)

Telencephalon   1
Cortex areas at the splenium of the corpus callosum. Borders on the isthmus of cingulate gyrus.

Retrotonsillar veins

Vessels   2
The retrotonsillar veins collect venous blood from the tonsil of cerebellum and transport it to the inferior hemisphere vein, via which blood flows into the straight sinus.

RF   1

Reticular formation

Rhinal sulcus

Telencephalon   2

TA Latin: *Sulcus rhinalis*
TA English: *Rhinal sulcus*
The entorhinal cortex (area 28) lies in this region. The transition between allocortex of the hippocampus and the cerebral cortex of the temporal lobe is to be found here.

Rhinencephalon
General CNS   3
The olfactory bulb and those structures that receive afferents form the olfactory bulb are classified as being part of the rhinencephalon. They include primarily the olfactory tract and the basal olfactory area, parts of the amygdaloid body, septum verum and prepiriform cortex.

Rhombencephalon
General CNS   3
TA Latin: *Rhombencephalon*
TA English: *Rhombencephalon*
Rhombencephalon is composed of myelencephalon, cerebellum and pons.

Rostrum of the corpus callosum
Telencephalon   1
TA Latin: *Rostrum corporis callosi*
TA English: *Rostrum of corpus callosum*
Beak-like extension of the corpus callosum at the genu of the corpus callosum.

Rubrobulbar tract
Mesencephalon   2
TA Latin: *Tractus rubrobulbaris*
TA English: *Rubrobulbar tract*
Descending fibers of the rubrobulbar tract and rubrospinal tract terminate on the interneurons in the lateral reticular formation and the dorsolateral intermediate zone of the spinal cord and directly on motoneurons of the nucleus of the facial nerve. The rubrobulbar tract and rubrospinal tract have a somatotopic arrangement.

Rubrospinal tract
Mesencephalon   3
TA Latin: *Tractus rubrospinalis*
TA English: *Rubrospinal tract*
Somatotopically arranged fiber bundles between the red nucleus and spinal cord. Runs in the lateral column of the spinal cord, originating in the magnocellular portion of the red nucleus, going to the spinal cord segments as far as the thoracic cord.
Regulates the tone of important flexors.

S

Saccades
Eye
→ Superior colliculus

Sacral pelvic foramina
Meninges & Cisterns   1
TA Latin: *Foramina sacralia ant.*
TA English: *Anterior sacral foramina*
A window in the sacral bone, opening in the direction of the pelvis.

Sacral segment of the spinal cord
Medulla spinalis   1
TA Latin: *Pars sacralis medullae spinalis*
TA English: *Sacral part of spinal cord*
Sacral cord. The segment of the spinal cord comprising the spinal nerves of the sacrum.

Sagittal layer
Telencephalon   3
Part of the optic radiation. Large fiber bundles from the lateral geniculate body run here to the area 17 (striate cortex) on the calcarine sulcus of the occipital lobe.

Salivatory nuclei
Pons   1
TA Latin: *Nuclei salivatorii*
TA English: *Salivatory nuclei*
There is an inferior and a superior salivatory nucleus. The fibers from the superior nucleus emerge with the intermediate nerve, while the fibers of the inferior salivatory nucleus leave the brainstem in the glossopharyngeal nerve and pass on to the parotid gland.

Salivatory nucleus
TA Latin: *Nucl. salivatorius*

TA English: *Salivatory nucleus*
→ Salivatory nuclei

Scotoma
Damage to the visual cortex of one hemisphere leads to anything from disruption of fields of vision (scotoma), directly correlated with the extent of damage, to homonymous hemianopsia (semi-blindness with disruption of one eye field).
If both visual cortices are affected, cortical blindness results. Eye reflexes such as pupillary reflex are preserved, but the cortex-related accommodation reflex is lost.

Secondary fissure
Cerebellum   2
TA Latin: *Fissura secunda*
TA English: *Secondary fissure*
The secondary fissure separates the tonsil of cerebellum from the biventer lobule.

Secondary motor cortex
Telencephalon   3
Is composed of the premotor cortex and the supplementary motor area. Involved in planning and initiating movements. Just before beginning an effector motor movement, a standby potential is generated here.

Secondary somatosensory cortex
Telencephalon   3
In the lower portion of the postcentral gyrus is situated a cytoarchitectonically slightly modified zone which reaches as far as the lateral sulcus and features a complete representation of the contralateral body half. This area is called the secondary somatosensory cortex, abbreviated SII.

Secondary visual cortex
Telencephalon   3
→ Area 18

Semicircular canals
Ear   2
TA Latin: *Canales semicirculares*
TA English: *Semicircular canals*

The semicircular canals of the vestibular system play a role in recording rotary movements and acceleration of the head. There is one semicircular canal for each spatial axis, hence totally three:
– anterior semicircular canal,
– lateral semicircular canal,
– posterior semicircular canal.

Semilunar gyrus

Telencephalon 2

The ambiens gyrus borders on the uncus and is partially surrounded by the semilunar gyrus. Both are components of the hippocampus.

Sensible

General CNS 3

Denotes "to the CNS". Hence sensory tracts conduct information from the periphery in the direction of the CNS.

Sensory

General CNS 3

"Sensory" also denotes fibers conducting information from the receptive fields of sensory organs in the direction of the CNS.

These include afferents from the retina, cochlea, olfactory epithelium and taste buds.

Sensory root of trigeminal nerve (V)

Nerves 3

TA Latin: *N. trigeminus (N.V), radix sensoria*
TA English: *Sensory root of trigeminal nerve (V)*

The sensory parts of the trigeminal nerve (V) arise from the spinal nucleus of the trigeminal nerve, pontine (principal) nucleus of the trigeminal nerve and mesencephalic nucleus of the trigeminal nerve. Together, they pass to the trigeminal ganglion and from here further in the three large branches: via the ophthalmic nerve (V1) innervate the orbita and forehead, via the maxillary nerve (V2) the nasal cavity, upper jaw and upper teeth, via the mandibular nerve (V3) the mandible, lower teeth and the anterior tongue.

Sensory tract

Pathways
→ **Medial lemniscus**

Septal nuclei

Telencephalon 2

TA Latin: *Nuclei septales*
TA English: *Septal nuclei*

The nuclei located in the septum verum (lateral septal nuclei, medial septal nuclei, nucleus of diagonal band) are involved in complex function circuits between hypothalamus and hippocampus. Lesions and stimulation show that the nuclei are involved in autonomic behavioral processes such as eating, drinking, micturition, defecation, sexual, reproduction and aggression behavior.

Septum pellucidum

Telencephalon 3

TA Latin: *Septum pellucidum*
TA English: *Septum pellucidum*

The septum pellucidum is a tissue lamina of glial cells spread out between the septum verum and corpus callosum. It separates the two lateral ventricles from each other and can also form a cavity between the two septal laminae, the cavum, also a liquid containing cavity.

Septum verum

Telencephalon 2

The septum verum is a region on the medial surface of the brain. It lies between the subcallosal area and the septum pellucidum.

In this area lie the septal nuclei which are an important component of function circuits between the hypothalamus and hippocampus and are involved in autonomic activities such as eating, drinking, micturition, defecation, sexual reproduction and aggression behavior.

Serotonergic cell groups B1-B8

1

TA Latin: *Cellulae serotoninergicae (B1-B8)*
TA English: *Serotoninergic cells (B1-B8)*

Neurons with serotonin as transmitter are found in the myelencephalon and mesencephalon, amidst the raphe nuclei accounting for between 15% and 70% of the cell mass.

Short ciliary nerves

Nerves 1

TA Latin: *Nn. ciliares breves*
TA English: *Short ciliary nerves*

From the ciliary ganglion isolated nerve fibers go to the eye and enter it via the sclera, where they provide parasympathetic innervation for the ciliary muscle (accommodation) and the sphincter of the pupil muscle (adaptation) as well sympathetic innervation for the dilator muscle of pupil.

Short gyri of insula

Telencephalon 2
TA Latin: *Gyri breves insulae*
TA English: *Short gyri of insula*
The function of this region is not really known, but viscerosensory and visceromotor functions are suspected.

Sigmoid sinus

Vessels 3
TA Latin: *Sinus sigmoideus*
TA English: *Sigmoid sinus*
The sigmoid sinus is the S-shaped continuation of the transverse sinus. It begins at the entry point of the superior petrosal sinus and stretches as far as the superior bulb of the jugular vein.
Taken up on its journey are: superior petrosal sinus, vestibular aqueductal vein and mastoid emissary veins.

Simple lobule

Cerebellum 2
TA Latin: *Lobulus simplex*
TA English: *Simple lobule*
The simple lobule belongs to the posterior lobe and is part of the cerebellar hemispheres. Apart from the areas in proximity to the vermis (intermediate part), the hemispheres belong to the phylogenetically young neocerebellum and receive their afferents via the mossy fibers of the pontocerebellar tract from the pontine nuclei. All hemisphere segments are hence also assigned to the pontocerebellum.

Sinus of the dura mater

Vessels 3
TA Latin: *Sinus durae matris*
TA English: *Dural venous sinuses*
The venous sinuses of the dura mater of the brain are called the sinuses of the dura mater.

They transport the venous blood from the brain, cranial meninges, and orbita to the jugular vein.

Sitomania and obesity

Hypothalamic nucleus that plays a decisive role in combat and satiation behavior (lesions mediate extreme anger outbursts, hyperphagia and obesity). Direct and indirect afferents from the subiculum and amygdaloid body, peripeduncular nucleus, lateral parabrachial nucleus as well as many other centers. Efferents go to the periaqueductal gray and to the reticular formation of mesencephalon.

Small cerebral vein (Galen)

Vessels
TA Latin: *V. interna cerebri*
TA English: *Internal cerebral vein*
→ Internal cerebral vein

Solitariospinal tract

Pathways 2
TA Latin: *Tractus solitariospinalis*
TA English: *Solitariospinal tract*
Efferents of the solitary nucleus passing directly to the spinal cord and probably implicated in autonomic reflexes.

Solitary nucleus

Myelencephalon 3
TA Latin: *Nucl. solitarius*
TA English: *Solitary nucleus*
Long cell column in the floor of the fourth ventricle, at the level of cranial nerves X, IX and VII. The nucleus has two parts:
– solitary nucleus, gustatory part
 Here terminate relevant fibers of cranial nerves X, IX and VII.
– solitary nucleus, cardiorespiratory part
Here terminate mucosa-innervating sensory fibers from cranial nerves VII, IX, and X.
Efferents go to the dorsal nucleus of the vagus nerve, medial parabrachial nucleus and to the dorsal tegmental nucleus. Direct fibers to the spinal cord course via the solitariospinal tract.

Solitary nucleus, cardiorespiratory part

Myelencephalon 2

In the caudal segment of the solitary nucleus terminate sensory fibers:
- of the vagus nerve (X) from the heart, mucosa of the airways and of the digestive tract,
- of the glossopharyngeal nerve (IX) from the middle ear mucosa, pharynx and oesophagus as well as
- of the facial nerve (VII) from the acoustic meatus and tongue.

→ Solitary tract

Solitary nucleus, gustatory part

Myelencephalon   2

In the rostral part of the solitary nucleus, the gustatory nucleus, terminate viscero-afferents from:
- the vagus nerve (X),
- the glossopharyngeal nerve (IX) and
- the facial nerve (VII).

→ Solitary tract

Solitary tract

Myelencephalon   2

TA Latin: *Tractus solitarius*
TA English: *Solitary tract*

The solitary tract comprises afferent fibers of cranial nerves VII, IX and X, which after entering the brainstem embark on a rostrocaudal course to gradually terminate in the solitary nucleus.

Somatic sensation

Lesions of the postcentral gyrus reduces the response to tactile, thermal and noci stimuli from the contralateral body half.

Somatotopic arrangement

General CNS   3

A somatotopic arrangement of a neuronal projection means that the projection areas occupy a relatively similar position vis-à-vis each other as that encountered in the original areas, from which they conduct information.

Examples of this are the "somatosensory homunculus" of the postcentral gyrus, or the equally strict somatotopic structure of the precentral gyrus (area 4).

Spastic torticollis

Damage to the corpus striatum results in the typically manifest symptoms of chorea, due to disinhibition of the globus pallidus and substantia nigra. Chorea is characterized at an advanced stage by hyperkinesia especially of the distal extremities' musculature and of the face. Dystonic syndrome (e.g. retrocollis, spastic torticollis) or athetosis are also encountered.

Sphenopalatine artery

Vessels   2

TA Latin: *A. sphenopalatina*
TA English: *Sphenopalatine artery*

Arises from the maxillary artery and passes through the sphenopalatine foramen into the posterior nasal cavity.

It supplies the mucosa of the nasal cavity and the nasal septum.

Sphenoparietal sinus

Vessels   2

TA Latin: *Sinus sphenoparietalis*
TA English: *Sphenoparietal sinus*

The sphenoparietal sinus generally emerges from the lateral branch of the superior sagittal sinus, receiving dural veins, diploic veins and the large superficial middle cerebral vein. From here the blood flows via the cavernous sinus and superior petrosal sinus into the sigmoid sinus.

Spinal branch of the intercostal vein

Vessels   1

TA Latin: *V. spinalis*
TA English: *Spinal vein*

The intercostal vein has lateral branches that penetrate the spinal root, collecting blood from here.

Spinal branch of the superior intercostal vein of the vermis

Vessels

→ Superior artery of vermis

Spinal cord

Medulla spinalis   3

TA Latin: *Medulla spinalis*
TA English: *Spinal cord*

Spinal cord is surrounded by the spinal meninges, enclosed in the vertebral canal and extends in adults to about the second lumbar vertebra. Here are found primarily conduction pathways, synaptic centers but also motor programs (simple and complex reflexes, movement programs, inter alia).

Spinal cord, cervical part

Medulla spinalis 1
TA Latin: *Medulla spinalis, pars cervicalis*
TA English: *Spinal cord, cervical part*
Cervical cord. The part of the spinal cord comprising the spinal nerves of the cervical vertebral column.

Spinal cord, gray matter

Medulla spinalis 3
TA Latin: *Medulla spinalis, Subst. grisea*
TA English: *Spinal cord, grey matter*
Here are found nuclear regions in which the fibers synapse.
Three areas are distinguished:
– posterior horn,
– intermediate substance,
– anterior horn.

Spinal cord, lumbar part

Medulla spinalis 1
TA Latin: *Medulla spinalis, pars lumbalis*
TA English: *Spinal cord, lumbar part*
Lumbar cord. The part of the spinal cord comprising the spinal nerves of the lumbar vertebral column.

Spinal cord, thoracic part

Medulla spinalis 1
TA Latin: *Medulla spinalis, pars thoracica*
TA English: *Spinal cord, thoracic part*
Thoracic cord. The part of the spinal cord comprising the spinal nerves of the thoracic vertebral column.

Spinal dura mater

Meninges & Cisterns 3
TA Latin: *Dura mater spinalis*
TA English: *Spinal dura mater*
Beneath the foramen magnum the periosteum and dura mater separate, giving rise to the epidural cavity, which spreads across the entire length of the vertebral column.
Cervical, thoracic and lumbar epidural cavities are also called peridural cavity.
The peridural cavity plays a decisive role in epidural anesthesia.

Spinal ganglion

Nerves 3
TA Latin: *Ganglion spinale*
TA English: *Spinal ganglion*
Spinal ganglion is formed by the cell nuclei of vicero-sensory, bi- or multipolar neurons, which project across the dorsal root into the spinal cord, where they mostly synapse directly on viscero-efferent fibers, thus creating the basis for autonomic reflex arcs.

Spinal nerve

Medulla spinalis 3
TA Latin: *N. spinalis*
TA English: *Spinal nerve*
A spinal nerve originates in the spinal cord, unlike the cranial nerve which arises from the cerebrum. A distinction is made between 31 pairs: 8 cervical (C1-C8), 12 thoracic (Th1–Th12), 5 lumbar (L1–L5), 5 sacral (S1–S5), and 1 coccygeal. Each segment pair provides sensory innervation for a clearly delineated skin area (=dermatome). Close to the spinal cord, the spinal nerve divides into a sensory dorsal root and a motor ventral root. External to the intervertebral foramen it divides again into a ventral branch and a dorsal branch.

Spinal nerves

Medulla spinalis 3
TA Latin: *Nn. spinales*
TA English: *Spinal nerves*
Each pair exits from the vertebral column, giving rise to 8 cervical, 12 thoracic, 5 lumbar, 5 sacral and one coccygeal spinal nerves, totally 31 pairs.

Spinal nucleus of the trigeminal nerve

Myelencephalon 2
TA Latin: *Nucl. spinalis n. trigemini*
TA English: *Spinal nucleus of trigeminal nerve*
This nucleus extends from the principal nucleus of the trigeminal nerve to the dorsal column of

the cervical cord, which it enters. Afferents are the axons from the trigeminal ganglion, which convey somatotopically organized impulses from the face via the spinal tract of the trigeminal nerve. Efferents come from the caudal nuclear region, decussate to the contralateral side and pass as the lateral trigeminothalamic tract to the ventral posteromedial thalamic nucleus.

Spinal nucleus of the trigeminal nerve, caudal part

Myelencephalon   2
TA Latin: *Nucl. spinalis n. trigemini, pars caudalis*
TA English: *Spinal nucleus of trigeminal nerve, caudal part*
The far-reaching spinal nucleus of the trigeminal nerve is organized somatotopically:

- oral part: perioral skin, as far as nose tip,
- interpolar part: middle of chin, via cheek bones and dorsal nose to above the eyebrows.
- caudal part: tip of chin, ear region, forehead, scalp.

Efferents of this nucleus pass here as the lateral trigeminothalamic tract to the contralateral thalamus.

Spinal nucleus of the trigeminal nerve, interpolar part

Myelencephalon   2
TA Latin: *Nucl. spinalis n. Trigemini, pars interpolaris*
TA English: *Spinal nucleus of trigeminal nerve, interpolar part*
→ Spinal nucleus of the trigeminal nerve, caudal part

Spinal nucleus of the trigeminal nerve, oral part

Myelencephalon   2
TA Latin: *Nucl. spinalis n. Trigemini, subnucleus oralis*
TA English: *Spinal nucleus of the trigeminal nerve, oral subnucleus*
→ Spinal nucleus of the trigeminal nerve, caudal part

Spinal root of the accessory nerve

Nerves
TA Latin: *Radix spinalis n. accessorii*

TA English: *Spinal root of accessory nerve*
→ Accessory nerve (XI)

Spinal roots in the dural sheath (between epidural veins)

Medulla spinalis   2
Roots and ganglia of the spinal nerves are also covered with dura mater. The latter is suspended to the intervertebral foramina via a connective tissue closure.

Spinal tract of the trigeminal nerve

Myelencephalon   2
TA Latin: *Tractus spinalis n. trigemini*
TA English: *Spinal tract of trigeminal nerve*
The spinal tract extends from the entry point of the trigeminal nerve at the level of the pons as far the lower end of the spinal nucleus of the trigeminal nerve. It carries the afferent fibers of this nucleus and its fibers are somatotopically organized and also project somatotopically:

- spinal nucleus of the trigeminal nerve, caudal part;
- spinal nucleus of the trigeminal nerve, interpolar part;
- spinal nucleus of the trigeminal nerve, oral part.

Spinal veins (anterior, lateral, posterior)

Vessels   3
TA Latin: *Vv. spinales (ant., lat., post.)*
TA English: *Spinal veins (anterior, lateral, posterior)*
Veins of the spinal cord. They collect venous blood from the spinal cord and spinal meninges and carry it into the internal vertebral venous plexus.

Spino-olivary tract

Medulla spinalis   2
TA Latin: *Tractus spinoolivaris*
TA English: *Spino-olivary tract*
Carries afferents from the spinal cord to the two accessory olives (dorsal accessory nucleus of the inferior olive and medial accessory nucleus of the inferior olive) and courses in the anterolateral fasciculus.

Spinocerebellum

Cerebellum 2
TA Latin: *Spinocerebellum*
TA English: *Spinocerebellum*
Phylogenetically, a very old part of the cerebellum. Corresponds to the vermis cerebelli with its surrounding intermediate part (paravermal part). The afferents of this region come from the spinal cord, hence this part is also called the spinocerebellum. Also called Palaeocerebellum.

Spinoreticular fibers

Pathways 1
TA Latin: *Fibrae spinoreticulares*
TA English: *Spinoreticular fibres*
Projections of the spinal cord to parts of the reticular formation.

Spinoreticular projection

Pathways
TA Latin: *Tractus spinoreticularis*
TA English: *Spinoreticular tract*
→Spinothalamic tract, Spinoreticular tract

Spinoreticular tract

Pathways 2
TA Latin: *Tractus spinoreticularis*
TA English: *Spinoreticular tract*
Projections of the intermediate substance of the spinal cord to neurons of the reticular formation. They continue from here in several steps to the intralaminar thalamic nucleus and the central gray matter of mesencephalon.

Spinotectal tract

Pathways 2
TA Latin: *Tractus spinotectalis*
TA English: *Spinotectal tract*
Conveys information from the spinal cord to the nuclei of the quadrigeminal plate (inferior colliculus and superior colliculus).
Runs in the anterolateral fasciculus.

Spinothalamic tract

Pathways 2
TA Latin: *Tractus spinothalamicus*
TA English: *Spinothalamic tract*
The sensory anterior column runs in the anterolateral fasciculus and conducts protopathic (pressure, temperature, pain, touch) information of the contralateral body half to the ventral posterolateral thalamic nucleus. Decussation to the contralateral side is effected segmentwise in the white commissure of the spinal cord. Having synapsed in the thalamus, information flow continues to the postcentral gyrus, the primary somatosensory cortex.

Spinous foramen

Meninges & Cisterns 1
TA Latin: *Foramen spinosum*
TA English: *Foramen spinosum*
Point of passage, situated in the great wing of sphenoid bone, for the middle meningeal artery and of some branches of the mandibular nerve into the middle cranial fossa.

Spinovestibular fibers

Medulla spinalis 2
Isolated projections from the spinal cord to vestibular nuclei.

Spiral organ

Ear 2
TA Latin: *Organum spirale*
TA English: *Spiral organ*
= Corti-Organ. Inner and outer hairy cells form the spiral organ.

Splenial artery

Vessels
TA Latin: *A. splenica*
TA English: *Splenic artery*
→ Pericallosal artery, posterior branch

Splenium of the corpus callosum

Telencephalon 3
TA Latin: *Splenium corporis callosi*
TA English: *Splenium of corpus callosum*
In the splenium of the corpus callosum course the fibers interconnecting the occipital lobes. The U-shaped course of the fibers is called the occipital forceps.

SSA (special somato-afferent series)

General CNS
→ CSA (common somato-afferent series)

Stellate cells

Telencephalon   3
The cortex consists of very different neuron types, which can be categorized into two main types: the pyramidal cells and star cells. About 80% of cortical neurons are pyramidal cells.

Stilling

Cerebellum   2
TA Latin: *Commissura cerebelli*
TA English: *Cerebellar commissure*
→ Cerebellar commissure

Straight sinus

Vessels   3
TA Latin: *Sinus rectus*
TA English: *Straight sinus*
The straight sinus collects the venous blood of the deep cerebral veins (e.g. great cerebral vein) in the region of the splenium of the corpus callosum, transporting it to the confluence of the sinuses, situated at the occipital pole of the cerebellum.
From here it flows to the jugular vein via the transverse sinus and sigmoid sinus.

Stratum cinereum

Mesencephalon
TA Latin: *Colliculus sup.,*
Stratum griseum superficiale
TA English: *Superior colliculus,*
superficial grey layer
→ Superior colliculus, superficial gray layer

Stria terminalis

Diencephalon   3
TA Latin: *Stria terminalis*
TA English: *Stria terminalis*
The stria terminalis is the most important efferent of the amygdaloid body.
It is a bundle of myelinated fibers coursing in the lateral ventricle, in the groove between thalamus and caudate nucleus and dividing at the anterior commissure. Target areas are: preoptic area, anterior hypothalamic area, hypothalamic nuclei, interstitial nucleus of stria terminalis. It marks the border between diencephalon and telencephalon.

Striate branches of the middle cerebral artery

Vessels
TA Latin: *A. cerebri media, Rr. striati*
TA English: *Middle cerebral artery, striate branches*
→ Middle cerebral artery, striate branches

Striate cortex

Telencephalon   3
→ Area 17 (striate cortex)

Striatonigral fibers

Pathways   1
Efferent fibers of the corpus striatum which project to the substantia nigra.

Striatum   3

TA Latin: *Striatum*
TA English: *Striatum*
→ Corpus striatum

Subarachnoid cisterns

Meninges & Cisterns   3
TA Latin: *Cisternae subarachnoideae*
TA English: *Subarachnoid cisterns*
The space between the arachnoid and pia mater is called the subarachnoid space. Since the arachnoid rests on the calvaria, while the pia mater rests on the surface of the brain, large subarachnoid spaces called cisterns, subarachnoid cisterns, are formed due to invaginations of the brain. The most important cisterns are the cerebellomedullary cistern, cisterna ambiens, interpeduncular cistern and chiasmatic cistern. The supracellary cisterns designate chiasmatic cistern + interpeduncular cistern + cistern of lamina terminalis.

Subarachnoid space

Meninges & Cisterns   3
TA Latin: *Spatium subarachnoideum*
TA English: *Subarachnoid space*
The subarachnoid space lies within the leptomeninx, i.e. between arachnoid and pia mater. It is filled with CSF and features an interspersed trabecular structure of spiderlike fine threads which open out between arachnoid and pia ma-

ter. The name (arachnos = spider) also derives from this fact.

Subcallosal area

Telencephalon   1
TA Latin: *Area subcallosa*
TA English: *Subcallosal area*
Area situated on the inside of the frontal lobe, at the end of the cingulate gyrus.

Subclavian artery

Vessels   1
TA Latin: *A. subclavia*
TA English: *Subclavian artery*
The left subclavian artery arises from the aortic arch, and the right from the brachiocephalic trunk.
It supplies parts of the head, brain, neck, spinal cord, breast and arm.

Subclavian vein

Vessels   3
TA Latin: *V. subclavia*
TA English: *Subclavian vein*
Transports venous blood from the arm, shoulder and lateral thoracic wall into the brachiocephalic vein.

Subcoerulean area (A6sc)

Mesencephalon   1
TA Latin: *Nucl. subcaeruleus*
TA English: *Subcaerulean nucleus*
Noradrenergic cells, which are summarized as group A6sc and also belong to the lateral reticular formation and the monoaminergic cell groups, are also to be found in an area beneath the locus coeruleus, the subcoerulean area.

Subcoerulean nucleus

Mesencephalon   1
TA Latin: *Nucl. subcaeruleus*
TA English: *Subcaerulean nucleus*
→ Subcoerulean area (A6sc)

Subcostal artery

Vessels   1
TA Latin: *A. subcostalis*
TA English: *Subcostal artery*

Arises from the thoracic aorta and is the lowest intercostal artery.

Subcostal vein

Vessels   2
TA Latin: *V. subcostalis*
TA English: *Subcostal vein*
Courses beneath the 12th rib and on the left enters the hemiazygos vein and on the right the azygos vein. The venous blood originates from the back muscles, vertebral canal and spinal column.

Subdural space

Meninges & Cisterns   3
TA Latin: *Spatium subdurale*
TA English: *Subdural space*
Spatium = space. The hypothetic space between dura mater and arachnoid (mater) is called subdural space. Generally the arachnoid is fit tightly and inseparable to the dura mater. Only bleedings or similar can result in separation of these two meninx (subdural hemorrhage).

Subfornical organ

Meninges & Cisterns   2
TA Latin: *Organum subfornicale*
TA English: *Subfornical organ*
A circumventricular organ situated at the interventricular foramen (of Monro). Found here are nerve endings with somatostatin and luliberin, the latter is involved in, inter alia, rhythmic release of FSH and LH. In the subfornical organ, angiotensin II enters the brain tissue where it triggers release of vasopressin by the supraoptic nucleus and the paraventricular nucleus.

Subiculum

Telencephalon   1
TA Latin: *Subiculum*
TA English: *Subiculum*
The subiculum is a band of cells that deep in the hippocampal sulcus continues the CA1 cell layer of Ammon´s horn and, for its part, joins the cell band of the presubiculum. It thus marks the transition from hippocampus to the area surrounding hippocampus.
In the subiculum most efferents arise from the hippocampus (→ fornix), afferents come from the entorhinal area primarily.

Sublenticular part of internal capsule

Telencephalon   2

TA Latin: *Capsula interna, pars sublentiformis*

TA English: *Sublentiform limb of internal capsule*

The portion of the internal capsule situated on the upper part of the tail of caudate nucleus, retrolenticular part contains, inter alia, the optic radiation as well as the occipitopontine tract.

The portion situated in the lower part, sublentiform part, conversely contains parts of the auditory radiation as well as the temporopontine tract.

Substantia gelatinosa of the spinal nucleus of the trigeminal nerve, caudal part

Medulla spinalis   2

Substantia gelatinosa (of Roland) at the level of the spinal nucleus of the trigeminal nerve, caudal part. Pain fibers synapse here.

Substantia gelatinosa (of Roland)

Medulla spinalis   2

TA Latin: *Subst. gelatinosa*

TA English: *Gelatinous substance*

A small-celled area in the posterior horn of the spinal cord. Pain fibers synapse here.

Substantia innominata

Telencephalon   1

TA Latin: *Subst. innominata*

TA English: *Innominate substance*

In the triangle of amygdaloid body, hypothalamus and putamen lies the substantia innominata in which the large and important basal nucleus of Meynert (Ch.4) is embedded.

Cells and arrangement are somewhat similar to those of the globus pallidus and putamen.

Substantia nigra

Mesencephalon   3

TA Latin: *Subst. nigra*

TA English: *Substantia nigra*

The substantia nigra is the largest nucleus of the mesencephalon. A distinction is made between:
– substantia nigra, pars compacta,
– substantia nigra, pars reticulata.

There is a very close, possibly even reciprocal point-to-point, connection between corpus striatum and substantia nigra.

The substantia nigra, pars compacta, plays an important role in Parkinson's disease.

Substantia nigra, pars compacta

Mesencephalon   2

TA Latin: *Subst. nigra, pars compacta*

TA English: *Substantia nigra, compact part*

The dorsal, large-celled segments of the substantia nigra are globally known as pars compacta. The large, polygonal and dopamine-producing cells lie close together.

Their very fine dopaminergic efferents form direct synaptic contacts with striatonigral projection neurons. This projection plays an important role in the initiation of voluntary, motor programs.

Parkinson's disease is characterized by progressive loss of neurons in the substantia nigra, pars compacta, degeneration of their ascending projections and reduction of the dopamine content in the corpus striatum. Symptoms include rigor, tremor, akinesia.

Substantia nigra, pars reticulata

Mesencephalon   2

TA Latin: *Subst. nigra, pars reticularis*

TA English: *Substantia nigra, reticular part*

The ventral, small-celled sections of the substantia nigra are called the pars reticulata. The cells (cell group A9) are less densely packed, and have a structure similar to that of the inner segment of the globus pallidus. They receive topically organized afferents from the caudate nucleus and globus pallidus (GABA, substance P, dynorphin, enkephalin). Efferents pass to the substantia nigra, globus pallidus, caudate nucleus, and putamen.

Subthalamic nucleus (Luys)

Diencephalon   3

TA Latin: *Nucl. subthalamicus (Luys)*

TA English: *Subthalamic nucleus (Luys)*

A well-circumscribed, large-celled nucleus in the caudalmost region of the diencephalon. It belongs with the globus pallidus, inter alia, to the subthalamus.

The lateral globus pallidus projects strictly topographically to the subthalamic nucleus, which in turn projects with inhibitory effects to all parts of the globus pallidus. Efferents also to the caudate nucleus, putamen and substantia nigra. GABA is the transmitter for both projections.

Subthalamus

Diencephalon   2
TA Latin: *Subthalamus*
TA English: *Subthalamus*
The subthalamus contains:
– globus pallidus,
– subthalamic nucleus,
– reticular area,
– ventral nucleus of the LGB,
– tegmental area H,
– zona incerta.
The region is an important synaptic center of the extrapyramidal, somatomotor system and belongs partly to the basal ganglia.

Sulcus limitans

Pons   1
TA Latin: *Sulcus limitans*
TA English: *Sulcus limitans*
Groove along the medial eminence on the bottom of the rhomboid fossa.

Sulcus of the corpus callosum

Telencephalon   2
TA Latin: *Sulcus corporis callosi*
TA English: *Sulcus of the corpus callosum*
Sulcus on the floor of the longitudinal fissure of cerebrum, directly on the roof of the corpus callosum. Here runs the medial longitudinal stria and the lateral longitudinal stria.

Superficial cerebral veins

Vessels   1
TA Latin: *Vv. superficiales cerebri*
TA English: *Superficial cerebral veins*
Collective designation for all superficial veins which convey their blood content into the superior sagittal sinus.
Their counterpart is the inferior cerebral veins that convey their blood content into the transverse sinus or the sphenoparietal sinus.

Superficial gray layer of the superior colliculus

Mesencephalon

TA Latin: *Stratum griseum superficiale colliculi sup.*
TA English: *Superficial grey layer of superior colliculus*
→ Superior colliculus, superficial gray layer

Superficial middle cerebral vein

Vessels   2
TA Latin: *V. media superficialis cerebri*
TA English: *Superficial middle cerebral vein*
The superficial middle cerebral vein collects the venous blood around the lateral sulcus and transports it to the cavernous sinus.
Via the superior anastomotic vein, it anastomoses with the superior sagittal sinus, while being connected via the inferior anastomotic vein with the transverse sinus.

Superficial temporal artery

Vessels   3
TA Latin: *A. temporalis superficialis*
TA English: *Superficial temporal artery*
A temporal artery coursing in the facial skin and supplying the skin and muscles of the lateral face.
Together with the maxillary artery, it emerges from the external carotid artery.

Superficial temporal veins

Vessels   1
TA Latin: *Vv. temporales superficiales*
TA English: *Superficial temporal veins*
Collect venous blood from the temporal region and carry it into the retromandibular vein.

Superior anastomotic vein (Trolard)

Vessels   2
TA Latin: *V. anastomotica sup. (Trolard)*
TA English: *Superior anastomotic vein (Trolard)*
The superior anastomotic vein connects the superior sagittal sinus with the superficial middle cerebral vein.
Hence together the superior and inferior anastomotic veins form an anastomosis between the superior sagittal sinus and the transverse sinus.

Superior and inferior choroid veins

Vessels 1

TA Latin: *Vv. choroidea (sup.+inf.)*

TA English: *Superior and inferior choroid veins*

The inferior choroid vein transports its venous blood from the choroid plexus of the lateral ventricle and hippocampus to the basal vein.

The superior choroid vein carries venous blood from the lateral choroid plexus of the lateral ventricle, from upper regions of the hippocampus and from the fornix and corpus callosum.

Blood flows via the internal cerebral vein into the great cerebral vein.

Superior artery of vermis

Vessels 1

TA Latin: *A. vermis sup.*

TA English: *Superior vermian artery*

Lateral branch of the posterior inferior cerebellar artery.

Courses via the uvula and along the cerebellar vermis, supplying the cortical regions of the cerebellum.

Superior bulb of jugular vein

Vessels 3

TA Latin: *Bulbus sup. v. jugularis*

TA English: *Superior bulb of jugular vein*

At the superior bulb of the jugular vein the sigmoid sinus joins the jugular vein which unites with the subclavian vein to form the brachiocephalic vein and then the superior vena cava.

Superior central nucleus (B6 + B8)

Mesencephalon 1

A serotonergic cell group near the central gray matter, which is a component of the limbic system.

Superior cerebellar artery

Vessels 3

TA Latin: *A. sup. cerebelli*

TA English: *Superior cerebellar artery*

The superior cerebellar artery is the largest cerebellar artery and branches off from the basilar artery, shortly before it divides.

It courses on the upper side of the cerebellum in the direction of the tuber vermis, where it divides into two branches:

– superior cerebellar artery, lateral branch
– superior cerebellar artery, medial branch

Superior cerebellar artery, lateral branch

Vessels 2

TA Latin: *A. sup. cerebelli, R. lat.*

TA English: *Superior cerebellar artery, lateral branch*

The lateral branch of the largest cerebellar artery supplies parts of the cerebellar hemisphere as well as the dentate nucleus of the cerebellum.

Superior cerebellar artery, medial branch

Vessels 2

TA Latin: *A. sup. cerebelli, R. med.*

TA English: *Superior cerebellar artery, medial branch*

The medial branch of the largest cerebellar artery supplies the dorsal regions of the cerebellar hemispheres, the superior cerebellar peduncle, the caudal colliculus, before joining with the superior artery of vermis.

Superior cerebellar artery, mesencephalic branch

Vessels 1

This branch is given off by the medial branch of the superior cerebellar artery and supplies the mesencephalon.

Superior cerebellar cistern

Meninges & Cisterns 3

TA Latin: *Cisterna pontocerebellaris*

TA English: *Pontocerebellar cistern*

The superior cerebellar cistern covers the surface of the cerebellum, thus filling the upper portion of the tentorium cerebelli. Anteriorly, it merges with the cistern ambiens and, inferiorly, with the cerebellomedullary cistern.

Superior cerebellar peduncle

Cerebellum 3

TA Latin: *Pedunculus cerebellaris sup.*

TA English: *Superior cerebellar peduncle*

The major cerebellar efferents pass through this peduncle: cerebellothalamic tract and cerebellorubral tract. Since they cannot be easily distinguished from each other they are collectively known as the "superior cerebellar peduncle".

The only afferent tract is the anterior spinocerebellar tract, conducting proprioceptive information from the spinal cord to the spinocerebellum.

Superior cerebellar peduncle, descending branch

Cerebellum 2

Shortly before reaching the red nucleus, a small portion of the cerebellar efferents branch downwards from the superior cerebellar peduncle. The branches are collaterals descending to the pons and myelencephalon, where they terminate in the reticular formation. Hence one of the many feedback loops, in which the cerebellum is integrated, is formed via the reticulocerebellar tract.

Superior cerebellar peduncle vein

Vessels 1

Vein coursing above the superior cerebellar peduncle, which connects the superior middle cerebellar vein with the lateral mesencephalic vein.

Superior cerebral pedunculus, ascending branch

Cerebellum 1

The superior cerebellar peduncle designates the large fiber bundle which leaves the cerebellum via the superior cerebellar peduncle, projecting in the direction of the red nucleus and ventral lateral thalamic nucleus. This bundle divides before reaching the red nucleus. The ascending part goes to the red nucleus and thalamus, while the superior cerebellar peduncle, descending branch descends as a collateral side arm to the reticular formation.

Superior cervical ganglion

Nerves 3

TA Latin: *Ganglion cervicale sup.*
TA English: *Superior cervical ganglion*
Sympathetic ganglion. The associated neurons are situated in the upper thoracic cord (the cervical cord has no sympathetic neurons).

Superior choroid vein

Vessels 2

TA Latin: *V. choroidea sup.*
TA English: *Superior choroid vein*
The superior choroid vein carries venous blood from the choroid plexus of the lateral ventricle, from the upper hippocampus as well as from the fornix and corpus callosum. The blood flows via the internal cerebral vein into the great cerebral vein.

Superior colliculus

Mesencephalon 3

TA Latin: *Colliculus sup.*
TA English: *Superior colliculus*
Upper hill of the quadrigeminal lamina. Involved in fast eye movements, synaptic center for optokinetic reflexes (saccades). Afferents from the retina and visual cortex (opticofacial winking reflex), inferior colliculus and auditory cortex (reflex movement in direction of source of noise). Involved in accommodation reflex. Efferents to oculomotor cranial nerve nuclei and spinal cord.
Damage to the superior colliculi results in interruption of reflex eye movements, but not in impairment of cognitive perceptivity (e.g. image recognition).

Superior colliculus, deep gray layer

Mesencephalon 1

TA Latin: *Colliculus sup., Stratum griseum profundum*
TA English: *Superior colliculus, deep grey layer*
The superior colliculus is subdivided into seven layers. The deep gray layer is the 6th layer.
In addition to small interneurons, this layer also features large efferent cells whose axons radiate into layer 7.

Superior colliculus, deep white layer

Mesencephalon 1

TA Latin: *Colliculus sup., Stratum medullare profundum*
TA English: *Superior colliculus, deep white layer*
The superior colliculus is subdivided into seven layers. The deep white layer is the 7th layer.
Afferents come from layer 4 (middle gray layer) and 6 (deep gray layer).
Efferents project as tectonuclear fibers to the tegmentum of mesencephalon.

Superior colliculus, lemniscal layer

Mesencephalon   1
TA Latin: *Colliculus sup., Stratum medullare intermedium*
TA English: *Superior colliculus, intermediate grey layer*
Also called middle white layer.
The superior colliculus is subdivided into seven layers. The lemniscal layer is the 5th layer.
Terminating here are the fibers of the spinotectal tract, occipital lobe as well as of the medial lemniscus and lateral lemniscus.

Superior colliculus, middle gray layer

Mesencephalon   1
TA Latin: *Colliculus sup., Stratum griseum intermedium*
TA English: *Superior colliculus, intermediate grey layer*
The superior colliculus is divided into seven layers. The middle gray layer is the 4th layer.
Afferents are the axons of the second neuron of the optic reflex pathway whose cell bodies are situated in the superficial gray layer (2nd layer). Efferents project across the 7th layer from the superior colliculus.

Superior colliculus, optic layer

Mesencephalon   1
TA Latin: *Colliculus sup., Stratum opticum*
TA English: *Superior colliculus, optic layer*
The superior colliculus is subdivided into seven layers. The optic layer is the 3rd layer. It is composed chiefly of afferent fibers.

Superior colliculus, superficial gray layer

Mesencephalon   1
TA Latin: *Colliculus sup., Stratum griseum superficiale*
TA English: *Superior colliculus, superficial grey layer*
Superior colliculus, superficial gray layer, also called stratum cinereum.
The superior colliculus is subdivided into seven layers. The superficial gray layer is the 2nd layer.
Afferents are the fibers of the optic tract, efferents project to layer 4 (middle gray layer).

Superior colliculus, zonal layer

Mesencephalon   1
TA Latin: *Colliculus sup., Stratum zonale*
TA English: *Superior colliculus, zonal layer*
The superior colliculus is subdivided into seven layers. The zonal layer is the 1st layer. It is composed of a few thin nerve fibers.

Superior edge of petrous part of temporal bone

Skeleton   1
Superior edge of petrous bone. The petrous bone is part of the temporal bone.

Superior frontal gyrus

Telencephalon   3
TA Latin: *Gyrus front. sup.*
TA English: *Superior frontal gyrus*
In the area of the frontal gyrus close to the precentral gyrus is situated the premotor cortex, which plays an important role in planning effector voluntary movements and has close interaction with the cerebellum, thalamic nuclei and basal ganglia.
At the level of the superior frontal gyrus is situated the frontal eye field, which is involved in planning voluntary eye movements.
Hyperactivity of these neurons due to hemorrhage or tumors causes conjugate movements of both eyeballs (déviation conjugée). Conversely, destruction of tissue causes ipsilateral déviation conjugée, since now the activity of the contralateral eye field no longer has an antagonist.

Superior frontal sulcus

Telencephalon   3
TA Latin: *Sulcus front. sup.*
TA English: *Superior frontal sulcus*
The superior frontal sulcus separates the medial frontal gyrus from the superior frontal gyrus lying above it.

Superior hemisphere vein

Vessels   2
TA Latin: *V. sup. cerebelli*
TA English: *Superior vein of cerebellar hemisphere*
Superficial veins drain blood from the cerebellar cortex and carry it into the tentorial sinus (con-

fluence of the infratentorial veins). The latter flows into the confluence of the sinuses.

Superior hypophyseal artery

Vessels   3

TA Latin: *A. hypophysialis sup.*

TA English: *Superior hypophysial artery*

Arises from the internal carotid artery, cerebral part, and supplies the hypophyseal stalk as well as the infundibulum and parts of the hypothalamus.

Superior intercostal artery

Vessels   1

TA Latin: *A. intercostalis suprema*

TA English: *Supreme intercostal artery*

Arises from the costocervical trunk and supplies intercostal spaces 1+2, deep cervical and back muscles as well as spinal cord and spinal meninges.

Superior intercostal vein

Vessels

TA Latin: *V. intercostalis sup.*

TA English: *Superior intercostal vein*

Veins coursing in the costal sulcus conducting blood from the back musculature, vertebral canal and vertebral column to the azygos vein and hemiazygos vein.

Superior intervertebral vein

Vessels

The intervertebral veins unite the hemiazygos vein flowing outside the vertebral canal with the internal vertebral venous plexus situated in the epidural cavity.

Superior longitudinal fasciculus

Pathways   2

TA Latin: *Fasciculus longitudinalis sup.*

TA English: *Superior longitudinal fasciculus*

With its two branches (anterior brachium and posterior brachium), the superior longitudinal fasciculus establishes connections between virtually all cortical areas. The part of the fasciculus connecting the motor (Broca's) speech center with the sensory (Wernicke's) speech center is called the arcuate fasciclulus.

Superior medullary velum

Cerebellum   3

TA Latin: *Velum medullare sup.*

TA English: *Superior medullary velum*

The superior medullary velum spreads out between the two superior cerebellar peduncles and in conjunction with these, forms the roof of the fourth ventricle. It supports the lingula cerebelli and connects the cerebellum with the quadrigeminal plate. From the latter emerge isolated fibers of the superior colliculus and inferior colliculus to the vermis cerebelli (tectocerebellar tract). They conduct visual and auditory information to the cerebellum.

Superior occipito-frontal fasciculus

Telencephalon   2

TA Latin: *Fasciculus occipitofrontalis sup.*

TA English: *Superior occipitofrontal fasciculus*

Association fibers connecting cortical areas of the frontal lobe with the insula as well as other temporal and occipital cortical areas.

Superior olivary complex

Mesencephalon   3

TA Latin: *Nucl. olivaris sup.*

TA English: *Superior olivary complex*

→ Superior olive

Superior olive

Mesencephalon   3

TA Latin: *Nucl. olivaris sup.*

TA English: *Superior olivary nucleus*

The superior olivary complex comprises the nuclei:

– nucleus of the trapezoid body,
– nucleus of the superior lateral olive,
– medial nucleus of the superior olive,

and is thus a vital synaptic center in the auditory tract, playing an important role in acoustic reflexes (reflex eye movements towards the source of noise, fright movements).

Superior ophthalmic vein

Vessels   1

TA Latin: *V. ophthalmica sup.*

TA English: *Superior ophthalmic vein*

Belongs to the group of the orbital veins. Collects venous blood from the eye, lid, lacrimal glands and eye muscles, orbita, and nasal cavity and carries it into the cavernous sinus.

Superior orbital fissure

Skeleton   1

TA Latin: *Fissura orbitalis sup.*
TA English: *Superior orbital fissure*
Bony margin of the base of the skull.

Superior parietal lobule

Telencephalon   3

TA Latin: *Lobulus parietalis sup.*
TA English: *Superior parietal lobule*
In the direction of the occipital pole, the inferior and superior lobules unite at the postcentral gyrus.

Analogous to the secondary motor cortex there is also a secondary sensory cortex for the somatosensory control; this is believed to stretch across both lobules and to be responsible for analysis, recognition and assessment of tactile information.

Superior parietal lobule, area 5 (SIII)

Telencephalon   2

Further processing of tactile stimuli appears to take place in this area. Lesions in this region render one unable to recognize objects touched, with the eyes closed (asterognosia).

Superior petrosal sinus

Vessels   3

TA Latin: *Sinus petrosus sup.*
TA English: *Superior petrosal sinus*
The superior petrosal sinus transports the venous blood from the cavernous sinus to the sigmoid sinus, via which it then drains off into the superior bulb of the jugular vein.

Superior sagittal sinus

Vessels   3

TA Latin: *Sinus sagittalis sup.*
TA English: *Superior sagittal sinus*
The superior sagittal sinus runs directly beneath the cranium at the base of the falx cerebri and passes from the frontal pole to the occipital pole, where it enters the confluence of the sinuses.
It anastomoses with the sphenoparietal sinus and transverse sinus, collecting blood from the superficial cerebral veins, i.e. all large, superficial cerebral veins.

Superior semilunar lobule

Cerebellum   2

TA Latin: *Lobulus semilunaris sup.*
TA English: *Superior semilunar lobule*
The superior semilunar lobule belongs to the posterior lobe and is part of the cerebellar hemispheres. Apart from the areas in proximity to the vermis (intermediate part), the hemispheres belong to the phylogenetically young neocerebellum and receive their afferents via the mossy fibers of the pontocerebellar tract from the pontine nuclei. All hemisphere segments are hence assigned to the pontocerebellum.

Superior temporal gyrus

Telencephalon   3

TA Latin: *Gyrus temporalis sup.*
TA English: *Superior temporal gyrus*
The superior temporal gyrus lies at the upper margin of the temporal lobe.
It comprises Broadmann areas 42 and 22, which together form the secondary auditory cortex, i.e. Wernicke's area. Here tones are processed associatively, hence interpreted, relativized and combined with memory contents.
Lesions in Wernicke's area cause sensory aphasia. Unlike motor aphasia, the patients are not capable of comprehending position, meaning of tones and noises. Neither can language be understood, nor can a motor noise with its potential hazard be detected. Musicality is also adversely affected.

Superior temporal sulcus

Telencephalon   3

TA Latin: *Sulcus temporalis sup.*
TA English: *Superior temporal sulcus*
The superior temporal sulcus separates the medial temporal gyrus from the superior temporal gyrus.

Superior thalamic branch

Vessels   1

→ Cingulothalamic artery, superior thalamic branch

Superior thalamic vein
Vessels 2
The superior thalamic vein is the largest vein of the thalamus. It emerges from the dorsal side of the thalamus. It emerges from the dorsal side of the thalamus and flows into the internal cerebral vein.

Superior thalamostriate vein
Vessels 1
TA Latin: *V. thalamostriata sup.*
TA English: *Superior thalamostriate vein*
Superior thalamostriate vein or also called terminal vein.
Carries venous blood from the posterior region of the frontal lobe and from the parietal lobe into the internal cerebral vein.

Superior vein of vermis
Vessels 2
TA Latin: *V. sup. vermis*
TA English: *Superior vein of vermis*
The superior vein of vermis runs along the upper portion of the vermis cerebelli, conveying its blood content either directly or via a detour into the straight sinus.

Superior vena cava
Vessels 3
TA Latin: *V. cava sup.*
TA English: *Superior vena cava*
Superior vena cava arises from the union of right and left brachiocephalic veins as well as of the azygos vein and flows into the right atrium of the heart.

Superior vermis branch
Vessels
→ Superior artery of vermis

Superior vestibular nucleus
Pons 2
TA Latin: *Nucl. vestibularis sup.*
TA English: *Superior vestibular nucleus*
→ Vestibular nuclei (medial, superior, inferior)

Suprachiasmatic nucleus
Diencephalon 3
TA Latin: *Nucl. suprachiasmaticus*
TA English: *Suprachiasmatic nucleus*

This nucleus belongs to the anterior group of the medial zone of the hypothalamus and lies directly above the optic chiasm. It plays a key role in circadian rhythms. Afferents come from the retina, raphe nuclei and subiculum. Efferents go to the medial preoptic nucleus and the posterior hypothalamus.

Supramammillary region
Mesencephalon 1
TA Latin: *Nucl. supramamillaris*
TA English: *Supramammillary nucleus*
Cells above the mammillary body.
Afferents from the lateral septal nucleus, lateral preoptic nucleus, interstitial nucleus of stria terminalis.
Projections to the hippocampus.

Supramarginal gyrus
Telencephalon 2
TA Latin: *Gyrus supramarginalis*
TA English: *Supramarginal gyrus*
The supramarginal gyrus is situated around the posterior branch of the lateral sulcus on both sides.
Functionally, it belongs to the extensions from the secondary somatosensory cortex, i.e. that cerebral cortex region involved in assessment, interpretation and processing of somatosensory stimuli.

Supraoptic nucleus
Diencephalon 3
TA Latin: *Nucl. supraopticus*
TA English: *Supra-optic nucleus*
Situated directly above the optic nerve, this nucleus together with the paraventricular nucleus forms the neuroendocrine system of the posterior lobe of the hypophysis. Efferents go to the neurohypophysis where they release into the blood ADH (vasopressin) and oxytocin. ADH (antidiuretic hormone) inhibits the permeability of renal epithelia to water. Oxytocin effects contraction of the uterus when giving birth and controls the release of milk during the lactation phase.
Dysfunction of this nuclear region induces the clinical manifestations of diabetes insipidus with severe polyuria (more than 20l per day),

since the lack of ADH results in a more or less unimpeded efflux of water from the renal epithelium.

Suprapineal recess

Meninges & Cisterns   2
TA Latin: *Recessus suprapinealis*
TA English: *Suprapineal recess*
At the habenular trigone, above the pineal body, the third ventricle curves dorsally. This evagination is called the suprapineal recess.

Suprasellar cisterns

Meninges & Cisterns   2
Chiasmatic cistern + interpeduncular cistern + terminal lamina cistern is called suprasellar cistern.

Supraspinal nucleus

Myelencephalon   1
TA Latin: *Nucl. supraspinalis*
TA English: *Supraspinal nucleus*
At the caudal end of the myelencephalon is the transition to the spinal cord. Encountered already at this juncture are motoneurons for the first cervical spinal nerves. These neurons belong to the supraspinal nucleus.

Supratrochlear artery

Vessels   2
TA Latin: *A. supratrochlearis*
TA English: *Supratrochlear artery*
Arises from the ophthalmic artery and supplies the forehead skin.

Surrounding area of hippocampus

Telencephalon   1
The hippocampus laterally joins a region characterized by the transition between allocortex and cerebral cortex. The following segments can be differentiated: CA1 of Ammon´s horn → subiculum (deep level of hippocampal sulcus) →presubiculum →parasubiculum → entorhinal area (all three in the parahippocampal gyrus) → perirhinal cortex (in the collateral sulcus).

SVA (special viscero-afferent series)

General CNS
→ CSA (common somato-afferent series)

SVE (special viscero-efferent series)

General CNS
→ CSA (common somato-afferent series)

Sympathetic branch (carotid plexus)

Nerves   1
On their journey from the superior cervical ganglion to the eye, the sympathetic fibers course parallel to the carotid artery as far as the carotid plexus. From here, they pass in the sympathetic branch to the ciliary ganglion, which they pass without synapsing, to then continue to the eye.

Sympathetic tone

→ Adrenergic cell groups C1,C2

T

Tail of caudate nucleus
Telencephalon 2
TA Latin: *Cauda nuclei caudati*
TA English: *Tail of caudate nucleus*
Rambling, c-shaped tail of caudate nucleus.

Tapetum
Telencephalon 1
TA Latin: *Tapetum*
TA English: *Tapetum*
Commissural fibers of the corpus callosum forming the roof of the lateral ventricle, inferior horn, and lateral ventricle, posterior horn, are called the tapetum.

Tectobulbar and tectospinal tract
Mesencephalon 2
TA Latin: *Tractus tectobulbaris + tectospinalis*
TA English: *Tectobulbar and tectospinal tract*
Deep layers of the superior colliculus project to nuclei of the brainstem (often called bulb). The greatest bundle descends close to the midline on the contralateral side, while one much smaller bundle continues on its ipsilateral course.
The tectospinal tract runs parallel over a long distance.

Tectocerebellar tract
Cerebellum 3
From the quadrigeminal plate, or more precisely from the inferior colliculi and the superior colliculi, fibers run directly via the superior medullary velum to central regions of the vermis cerebelli. These fibers conduct optic and auditory signals, with their projection fields overlapping with those of the trigeminocerebellar tract.

Tectopontine tract
Pathways 2
TA Latin: *Tractus tectopontinus*
TA English: *Tectopontine tract*
Via this bundle impulses from the visual system are passed on to the nuclei of the pons (Varolius), from where they are projected to the cerebellum.

Tectospinal tract
Mesencephalon 2
TA Latin: *Tractus tectospinalis*
TA English: *Tectospinal tract*
A large projection from the substantia nigra terminates in a very regular pattern of bands in the superior colliculus, middle gray layer. This layer contains the cells giving rise to the predorsal fasciculus, a major descending bundle, which dispatches a large number of fibers to premotor and motor centers in the brainstem and spinal cord. Runs in the medial longitudinal fasciculus.

Tectum of mesencephalon
Mesencephalon 3
TA Latin: *Tectum mesencephali*
TA English: *Tectum of midbrain*
→ Quadrigeminal plate

Tegmental area, (Forel's field)
Diencephalon 2
TA Latin: *Nuclei campi perizonalis (Forel-Feld)*
TA English: *Nuclei of perizonal fields (Forel's field)*
Between the thalamus and zona incerta is situated Forel´s tegmental field. It belongs to the subthalamus and consists of myelinated axons which belong to the thalamic fasciculus. This is composed of the lenticular fasciculus and ansa lenticularis and is the most important connection between the globus pallidus and ventral lateral thalamic nucleus.

Tegmental area H
TA Latin: *Nucl. campi med. H*
TA English: *Nucleus of medial field H*
→ Tegmental area, (Forel's field)

Tegmental pontine reticular nucleus (Bechterew)

Mesencephalon   1
TA Latin: *Nucl. reticularis tegmenti pontis*
TA English: *Reticulotegmental nucleus*
Belongs to the medial reticular formation.
It is a target region of the mammillotegmental tract and has major efferents to the cerebellum and pontine nuclei.

Tegmentum

3

TA Latin: *Tegmentum*
TA English: *Tegmentum*
→ Tegmental area, (Forel's field)

Tegmentum of mesencephalon

Mesencephalon   3
TA Latin: *Tegmentum mesencephali*
TA English: *Tegmentum of midbrain*
Quadrigeminal plate. This deep layer of the mesencephalon stretches between the tectum of mesencephalon and the cerebral peduncles, arising to the surface only deep in the interpeduncular fossa.
A number of cranial nerve nuclei are encountered here as well as the reticular formation of mesencephalon with its nuclei, the substantia nigra and parts of the central gray matter of mesencephalon.

Tegmentum of pons

Pons   1
TA Latin: *Tegmentum pontis*
TA English: *Tegmentum of pons*
The tegmentum of pons comprises the dorsal part of the pons and contains the nuclei of the cranial nerves: trigeminal nerve (V), abducens nerve (VI), facial nerve (VII) and vestibulocochlear nerve (VIII).

Telencephalon

Telencephalon   3
TA Latin: *Telencephalon*
TA English: *Telencephalon*
At a deep level, it is composed of the commissures, the limbic system and the basal ganglia, on the surface of both hemispheres it is composed of the greatly folded cortex (cerebral cortex). All "higher" brain functions are found here, e.g. voluntary motor control, motor and sensory speech, cognition, visual and auditory systems, surface and proprioceptive sensibility. Also memory as well as affective and emotional mechanisms.

Temporal lobe

Telencephalon   3
TA Latin: *Lobus temporalis*
TA English: *Temporal lobe*
Temporal lobe stretches from the temporal pole to the lateral sulcus.

Temporal operculum

Telencephalon   2
TA Latin: *Operculum temporale*
TA English: *Temporal operculum*
Operculum = lid.
The temporal operculum designates the part of the temporal lobe covering the insula, which is located deep in the lateral sulcus.

Temporal plane

Telencephalon   3
TA Latin: *Planum temporale*
TA English: *Temporal plane*
The floor of the lateral sulcus is formed by the upper side of the temporal lobe. This almost flat plane is called the temporal plane. It has characteristic, transverse gyri (transverse temporal gyrus) which are called Heschl's transverse convolutions. In this cortex area (area 41 and 42) terminates the auditory tract, hence the term auditory cortex or primary auditory cortex. This area is tonotopically organized and has large efferents in the surrounding auditory cortex.

Temporal polar cortex

Telencephalon   1
Cortical areas at temporal pole.

Temporal pole

Telencephalon   2
TA Latin: *Polus temporalis*
TA English: *Temporal pole*
Temporal pole of the brain located on the temporal lobe.

Temporal trunk
Vessels
→ Middle cerebral artery, posterior trunk

Temporo-ammonic tract or "perforatant path"
Telencephalon 1
Projections of the entorhinal cortex (area 28) to Ammon´s horn. The fibers enter the dentate gyrus.

Temporo-occipital artery
Vessels 2
TA Latin: *R. temporooccipitalis a. cerebri med.*
TA English: *Temporo-occipital branch of a. cerebri med.*
Arises from the middle cerebral artery, posterior trunk. The middle cerebral artery emerges from the internal carotid artery.
Passes via the temporal lobe deep into the occipital lobe, largely supplying its surface.

Temporopolar artery
Vessels 2
TA Latin: *A. polaris temporalis*
TA English: *Polar temporal artery*
Arises from the middle cerebral artery, posterior trunk. The middle cerebral artery emerges from the internal carotid artery.
Passes to the temporal pole, which it also supplies.

Temporopontine tract
Pathways 2
Projections with mostly motor information from the temporal lobe to the nuclei of the pons (Varolius)

Tenia
Meninges & Cisterns 2
TA Latin: *Taenia*
TA English: *Taenia*
The torn edge created on removing the choroid plexus or opening a ventricle is called tenia.

Tenia of the fornix
Meninges & Cisterns 2
TA Latin: *Taenia fornicis*
TA English: *Taenia of fornix*

Via the tenia of the fornix, the choroid plexus of the lateral ventricle is attached to the fornix. Parallel to it stretches the tenia of the fornix. In between, is a membrane (choroid tela of the lateral ventricle) supporting the choroid plexus of the lateral ventricle.

Tenia of the fourth ventricle
Pons 2
A torn edge is called a tenia. If for instance one removes the cerebellum from the brainstem, one separates the wall of the fourth ventricle. The torn edge thus created is called the tenia of the fourth ventricle.

Tenia of the thalamus
Meninges & Cisterns 2
TA Latin: *Taenia thalami*
TA English: *Taenia thalami*
By means of the tenia of the thalamus, the third ventricle is secured to the thalamus (stria medullaris). It joins the choroid tenia of the lateral ventricle.

Tentorial incision
Meninges & Cisterns 1
TA Latin: *Incisura tentorii*
TA English: *Tentorial notch*
The tentorium cerebelli spreads out here.

Tentorial sinus
Vessels 3
The tentorial sinus collects venous blood from the region of the tentorium cerebelli.
The confluence of the infratentorial veins and the confluence of the supratentorial veins can be differentiated. One collects the venous blood within the tentorium, whereas its counterpart collects the blood from the tectum of the latter.

Tentorium cerebelli
Meninges & Cisterns 3
TA Latin: *Tentorium cerebelli*
TA English: *Tentorium cerebelli*
The tentorium cerebelli is a layer of dura mater spreading between the cerebrum and cerebellum, between the edges of the petrous bone pyramid. The straight sinus stretches at the level of the tentorium cerebelli.

Terminal filament

Meninges & Cisterns   3
TA Latin: *Filum terminale*
TA English: *Filum terminale*
→ Filum terminale

Terminal nuclei

Diencephalon   2
Three nuclei of the diencephalon where fibers of the accessory optic tract terminate:
– dorsal terminal nucleus,
– lateral terminal nucleus,
– medial terminal nucleus.
All three play a role in the smooth pursuit of the eye as well as in visual-vestibulo interactions (coordination of eye and head movement etc.).

Terminal vein

Vessels
TA Latin: *V. thalamostriata sup.*
TA English: *Superior thalamostriate vein*
→ Superior thalamostriate vein

Thalamic branch (anteroinferior)

Vessels
→ Posterior communicating artery, anteroinferior thalamic branch

Thalamic fasciculus

Pathways   3
TA Latin: *Fasciculus thalamicus*
TA English: *Thalamic fasciculus*
The pallidothalamic projection is quantitatively the most important efferent system of the globus pallidus. It originates in the medial pallidal segment. Its fibers form first of all the lenticular fasciculus and the ansa lenticularis. The fibers pass into Forel´s field and, together, from here as the thalamic fasciculus to the rostral area of the ventral lateral thalamic nucleus. This projects to premotor cortex (area 6) (supplementary motor area). The rubrothalamic tract is likewise part of the thalamic fasciculus.

Thalamic nuclei

Diencephalon   1
TA Latin: *Nuclei thalami*
TA English: *Thalamic nuclei*

The thalamus is a nuclear complex in the wall of the diencephalon. By means of the internal medullary lamina, it is divided into three nuclear regions:
– ventral thalamic nuclei,
– lateral thalamic nuclei,
– median thalamic nuclei.

Thalamic pulvinar

Diencephalon   2
TA Latin: *Pulvinar thalami*
TA English: *Pulvinar*
The pulvinar belongs to the cell group of the thalamus and is composed of three parts: anterior part, medial part and inferior part.
The thalamic pulvinar receives afferents from the LGB and superior colliculus and projects primarily to the occipital, temporal and parietal cortex. In the frontal cortex it reaches only the eye field. Its tasks are not clear (speech, somatosensory interpretation).

Thalamic pulvinar, anterior part

Diencephalon   2
TA Latin: *Nucleus pulvinaris ant.*
TA English: *Anterior pulvinar nucleus*
The anterior cell groups of the thalamic pulvinar project together with the lateral posterior nucleus to area 5 and area 7 of the parietal lobe. These areas belong to the secondary somatosensory cortex and may be implicated in the interpretation of tactile stimulation.

Thalamic pulvinar, inferior part

Diencephalon   2
TA Latin: *Nucleus pulvinaris inf.*
TA English: *Inferior pulvinar nucleus*
The cells located in the lower segment of the pulvinar project to the infratemporal, visual association cortex and receive afferents from the superficial layers of the superior colliculus and the pretectal area.

Thalamic pulvinar, lateral part

Diencephalon   2
TA Latin: *Nucleus pulvinaris lat.*
TA English: *Lateral pulvinar nucleus*
The lateral nuclear groups of the thalamic pulvinar have widespread reciprocal connections with the association cortex in the posterior

parietal, occipital and temporal cortex. The connections appear to play a role in language formation, as they project to important speech centers. Afferents come from the superior colliculus and pretectal area.

Thalamic pulvinar, medial part

Diencephalon   2
TA Latin: *Nucleus pulvinaris med.*
TA English: *Medial pulvinar nucleus*
The median cell clusters of the thalamic pulvinar project into the superior temporal gyrus and to the temporal pole. Separate groups project to the frontal eye field.
Brodmann areas 42 and 22, i.e. the secondary auditory cortex (Wernicke's area) are located in the superior temporal gyrus.
Afferents come, inter alia, from the nonvisual layers of the superior colliculus.

Thalamic reticular nucleus

Diencephalon   1
TA Latin: *Nucl. reticularis thalami*
TA English: *Reticular nucleus of thalamus*
The thalamic reticular nucleus belongs to the subthalamus and is more like a lamina than a nucleus. It ascends from the lateral end of the zona incerta between the thalamus and internal capsule. Its intrinsic networks, lack of cortical projection as well as the transmitter GABA indicate that it is involved in postsynaptic inhibition, generally ensuing after stimulation of thalamic relay cells.

Thalamus

Diencephalon   3
TA Latin: *Thalamus*
TA English: *Thalamus*
The thalamus is a nuclear region in the wall of the diencephalon, on the floor of the lateral ventricle, central part. It is subdivided into groups by various fiber bundles (medullary laminae).
Functionally, the thalamus it the last computational, intergrational and relay station for sensory information on its way to the cerebrum. All sections of the cerebrum have reciprocal connections with the thalamus.

Thalamus, periventricular region

Diencephalon   1

TA Latin: *Nucl. med. thalami*
TA English: *Median nuclei of thalamus*
The medial thalamic nucleus is called the periventricular nuclear region of the thalamus, since it forms a layer containing nerve cells, beneath the ventricle ependyma.

Third ventricle

Meninges & Cisterns   3
TA Latin: *Ventriculus tertius*
TA English: *Third ventricle*
The third ventricle lies in the center of the diencephalon. Via an interventricular foramen, it is connected in each case with one lateral ventricle and conveys its CSF on to the fourth ventricle via the mesencephalic aqueduct.

Thoracic duct

Meninges & Cisterns   3
TA Latin: *Ductus thoracicus*
TA English: *Thoracic duct*
A lymph trunk formed from the fusion of three large lymph vessels. It collects lymph from the entire lower and upper left halves of the body and enters the venous angle.

Thoracic nucleus

Medulla spinalis   2
TA Latin: *Nucl. thoracicus post.*
(Stilling-Clarke)
TA English: *Posterior thoracic nucleus*
(Stilling-Clarke)
Cluster of ganglion cells near the dorsal column between segments C7 and L3. The axons form the posterior spinocerebellar tract and conduct proprio- and exteroceptive impulses of the lower limbs.

Tonsil of cerebellum

Cerebellum   2
TA Latin: *Tonsilla cerebelli*
TA English: *Tonsil of cerebellum*
The part of the hemisphere resting on the inferior medullary velum belongs to the posterior lobe.
Apart from the regions close to the vermis (intermediate part), the hemispheres belong to the phylogenetically young neocerebellum and receive their afferents via the mossy fibers of the pontocerebellar tract from the pontine nuclei.

Hence all hemispheres are summarized as also being part of the pontocerebellum.

Transverse fibers of the pons
Pathways
TA Latin: *Fibrae pontocerebellares*
TA English: *Pontocerebellar fibres*
→ **Pontocerebellar fibers**

Transverse pontine veins
Vessels   2
TA Latin: *Vv. pontis transversae*
TA English: *Transverse pontine veins*
Veins coursing obliquely above the pons. They belong to an irregular venous plexus around the pons, called the pontine veins.

Transverse process
Skeleton   1
TA Latin: *Processus transversus*
TA English: *Transverse process*
Transverse process of the vertebra.

Transverse sinus
Vessels   3
TA Latin: *Sinus transversus*
TA English: *Transverse sinus*
The large transverse sinus is arranged in pairs and transports the venous blood from the confluence of the sinuses to the sigmoid sinus. It has a diameter of around 1 cm and runs along the lateral edge of the tentorium.
On its path it receives the tentorial sinus.

Transverse sinus (lateral end)
Vessels   1
TA Latin: *Sinus transversus*
TA English: *Transverse sinus (lateral end)*
At the lateral end the transverse sinus enters the sigmoid sinus. The superior petrosal sinus also enters at this juncture.
From here, it enters the sigmoid sinus which enters into the internal jugular vein at the superior bulb of the jugular vein. The internal jugular vein transports the venous blood to the brachiocephalic vein, which in turn joins the superior vena cava.

Transverse sinus (medial end)
Vessels   1
TA Latin: *Sinus transversus*
TA English: *Transverse sinus (medial end)*
At its medial end, the transverse sinus is connected with the confluence of the sinuses, from which it receives its principle inflow of venous blood.

Transverse sinus of dura mater
Vessels
TA Latin: *Sinus durae matris*
TA English: *Dural venous sinuses*
→ **Sinus of the dura mater**

Transverse temporal gyri
Telencephalon
TA Latin: *Gyri temporales transversi (Heschl)*
TA English: *Transverse temporal gyri*
→ **Transverse temporal gyrus (Heschl)**

Transverse temporal gyrus (Heschl)
Telencephalon
TA Latin: *Gyrus temporalis transversus*
TA English: *Transverse temporal gyrus*
The floor of the lateral sulcus is formed by the upper side of the temporal lobe. This almost flat plane is called the temporal plane. It has characteristic, transverse gyri (transverse temporal gyrus (Heschl) which are called Heschl´s transverse convolutions. In this cortex area (area 41 and 42) terminates the auditory tract, hence the term auditory cortex or primary auditory cortex. This area is tonotopically organized and has large efferents in the surrounding auditory cortex.

Trapezoid body
Myelencephalon   3
TA Latin: *Corpus trapezoideum*
TA English: *Trapezoid body*
Efferents originating in the ventral cochlear nucleus decussate in a broad fiber band to the contralateral side. As soon as its reaches the contralateral side this band, the trapezoid body, kinks upwards in a 90° angle, and is now called the lateral lemniscus.
As such, it passes on to the inferior colliculus.

Trigeminal cavity

Meninges & Cisterns 2

TA Latin: *Cavum trigeminale*
TA English: *Trigeminal cave*

The trigeminal cavity is part of the pontocerebellar cistern. The dura mater forms a cavity in which is situated the trigeminal ganglion. This cavity contains CSF and is called the trigeminal cavity (or Meckel's Cave).

Trigeminal cistern

Meninges & Cisterns 1

The trigeminal cistern surrounds the trigeminal ganglion.

Trigeminal ganglion (Gasseri)

Nerves 3

TA Latin: *Ganglion trigeminale (Gasseri)*
TA English: *Trigeminal ganglion* (Gasseri)

In the middle of the cranial fossa is situated the ganglion of the trigeminal nerve (V). Here autonomic fibers synapse and here the nerve divides into its three major branches: ophthalmic nerve (V1), maxillary nerve (V2) and mandibular nerve (V3).

Trigeminal lemniscus

Mesencephalon 2

TA Latin: *Lemniscus trigeminalis*
TA English: *Trigeminal lemniscus*

Once the afferents have left the principle nucleus of the trigeminal nerve, they cross to the contralateral side as the ventral tegmental fasciculus, and form bundles here called the trigeminal lemniscus. The latter then passes together with the medial lemniscus and the spinothalamic tract to the thalamus.

Trigeminal nerve (V)

Nerves 3

TA Latin: *N. trigeminus (N.V)*
TA English: *Trigeminal nerve (V)*

The trigeminal nerve is the largest cranial nerve and has sensory/somatomotor functions. It provides sensory innervation for the entire face as well as for large parts of the cranial meninges and mucosa.

It provides motor innervation for the masticatory muscles.It emerges laterally from the pons, passes to the petrous bone, forming the tri-

geminal ganglion there in a dural fold. From here, three branches continue further: ophthalmic nerve (V1), maxillary nerve (V2) and mandibular nerve (V3)

Trigeminocerebellar tract

Cerebellum 3

Direct and indirect projections are thought to exit via the superior and inferior cerebellar peduncles from various trigeminal nuclear regions, conducting proprioceptive information from the face into the cerebellar hemispheres. The projection area of these mossy fibers overlap partially with the optic and auditory projection areas.

Trigeminothalamic tract

Mesencephalon 2

TA Latin: *Tractus trigeminothalamicus*
TA English: *Trigeminothalamic tract*

A distinction is made between three large trigeminothalamic connections. The dorsal trigeminothalamic tract passes on to the ipsilateral thalamus, and has its source in the principle nucleus of the trigeminal nerve. The trigeminal lemniscus connects the same nucleus with the contralateral thalamus. The lateral trigeminothalamic tract connects the spinal nucleus of the trigeminal nerve with the contralateral thalamus.

Trigone of the hypoglossal nerve

1

TA Latin: *Trigonum n. hypoglossi*
TA English: *Hypoglossal trigone*

Area of the fourth ventricle above the nucleus of the hypoglossal nerve, nuclear region of hypoglossal nerve (XII).

Trigone of the vagus nerve

Myelencephalon 2

TA Latin: *Trigonum n. vagi*
TA English: *Vagal trigone*

Area of the fourth ventricle above the dorsal nucleus of the vagus nerve, cranial nerve X.

Trochlear nerve (IV)

Nerves 3

TA Latin: *N. trochlearis (N.IV)*
TA English: *Trochlear nerve (IV)*

The trochlear nerve is the thinnest cranial nerve and is purely somatomotor in nature. It emerges as the sole cranial nerve on the dorsal side of the brainstem, beneath the inferior colliculus, and innervates only one muscle, the superior oblique muscle of the eyeball. By a change in tension, contraction of this muscle generates eye movement, laterally and downwards. The nerve reaches the orbita together with the oculomotor nerve (III) and abducens nerve (VI).
Skull: superior orbital fissure.

Trochlear nucleus
TA Latin: *Nucl. n. trochlearis*
TA English: *Nucleus of trochlear nerve*
→ Nucleus of the trochlear nerve

Trunk (body) of the corpus callosum
Telencephalon   3
TA Latin: *Truncus corporis callosi*
TA English: *Trunk of corpus callosum*
The trunk (body) of the corpus callosum is formed by the fibers interconnecting the temporal lobe and parietal lobe.

Trunk of the 8th thoracic nerve
Nerves   2
The fibers from the ventral root and the dorsal root of the thoracic cord together enter the intervertebral foramen as the trunk of the thoracic nerve.

Tuber cinereum
Diencephalon   2
TA Latin: *Tuber cinereum*
TA English: *Tuber cinereum*
The tuber cinerum forms the floor of the hypothalamus, giving rise to the infundibulum whose various segments form the neurohypophysis. This region contains the tuberal nucleus which plays an important role in regulation of the adenohypophysis (releasing factors).

Tuber vermis
Cerebellum   3
TA Latin: *Tuber vermis (VII B)*
TA English: *Tuber vermis (VII B)*
Segment of the vermis cerebelli lying below the folium vermis.

Like the entire vermis cerebelli, the tuber vermis also receives its afferents mainly from the spinal cord. It is thus part of the spinocerebellum = palaeocerebellum.

Tuberal nuclei
Diencephalon   1
TA Latin: *Nuclei tuberales*
TA English: *Tuberal nuclei*
The tuberal nuclei is the collective designation for the nuclei of the tuber cinereum. They include ventromedial hypothalamic nucleus and infundibular nucleus.

Tuberculum of the cuneate nucleus
Myelencephalon   3
Tuberculum on the dorsal transition from spinal cord to myelencephalon. The cuneate nucleus, in which the epicritic fibers from the upper extremities and neck region synapse, and which on the contralateral side become an important part of the medial lemniscus.

Tuberculum of the gracile nucleus
Myelencephalon   3
Tuberculum on the dorsal transition from spinal cord to myelencephalon. The gracile nucleus, in which the epicritic fibers from the lower extremities and trunk region synapse with the internal arcuate fibers, which on the contralateral side become an important part of the medial lemniscus.

Tuberoinfundibular tract
Diencephalon   1
Bundles of axons whose cell bodies lie in the infundibular nucleus. The axons course to the median eminence, ending on capillaries of the portal hypophyseal system. In the axons vesicles with releasing and release-inhibiting factors are transported, to be released in the capillaries for regulation of hormone secretion of the posterior lobe of the hypophysis.

Tuberothalamic artery
Vessels
TA Latin: *A. communicans post.,*
R. hypothalamicus

TA English: *Posterior communicating artery, hypothalamic branch*

→ **Posterior communicating artery, hypothalamic branch**

U

Uncal notch

Telencephalon   1

The uncal notch separates the uncus from the hippocampal gyrus and the gyrus ambiens.

Uncinate fasciculus of cerebellum

Cerebellum   3

TA Latin: *Fasciculus uncinatus cerebelli*
TA English: *Uncinate fasciculus of cerebellum*

The uncinate fasciculus of cerebellum represents the sum of all efferents to the fastigial nucleus.

It decussates to the contralateral side immediately after exiting from the nuclear region, dividing into two parts: the much larger part descends to the vestibular nuclei, while the smaller part ascends to the thalamus and hypothalamus (uncinate fasciculus of cerebellum, ascending branch).

Uncinate fasciculus of cerebellum, ascending branch

Cerebellum   1

Small fiber part of the uncinate fasciculus of cerebellum, which after decussation to the contralateral side, ascends to thalamic and hypothalamic nuclear regions.

Uncinate fasciculus of cerebrum

Telencephalon   2

TA Latin: *Fasciculus uncinatus cerebri*
TA English: *Uncinate fasciculus of cerebrum*

Association fibers connecting basal segments of the frontal lobe with the amygdaloid body, parahippocampal gyrus and parts of the temporal lobe.

Uncinate gyrus

Telencephalon   2

Part of Ammon´s horn and hence a vital part of the hippocampus, limbic system and memory formation.

Uncus

Telencephalon   2

TA Latin: *Uncus*
TA English: *Uncus*

The uncus, also called hippocampal uncus comprises:
– intralimbic gyrus,
– Giancomini's band,
– uncinate gyrus.

Giancomini's band belongs to the dentate gyrus, the other two to Ammon´s horn.

It is separated from the hippocampal gyrus and ambiens gyrus by the uncal notch.

Unimodal cortical association areas

Telencephalon   1

Association areas specializing in cognitive and associative processing of one sensory modality (e.g. area 18).

Uvula vermis

Cerebellum   3

TA Latin: *Uvula vermis (IX)*
TA English: *Uvula of vermis (IX)*

Segment of vermis cerebelli lying between the tonsil of cerebellum.

Like the entire vermis cerebelli, the uvula vermis also receives its afferents primarily from the spinal cord. It is thus part of the spinocerebellum = palaeocerebellum.

V

Vagus nerve, motor fibers

Nerves   1

Motor components of the vagus nerve (X). The somatomotor components come from nucleus ambiguus, the visceromotor from the dorsal nucleus of the vagus nerve.

Vagus nerve, sensory fibers

Nerves   1

Sensory components of the vagus nerve (X). The somatosensory fibers pass on to the spinal nucleus of the trigeminal nerve, the viscerosensory to the solitary nucleus.

Vagus nerve (X)

Nerves   3

TA Latin: *N. vagus (N.X)*
TA English: *Vagus nerve (X)*

Vagus nerve (X) is the largest parasympathetic nerve in the body, stretching to the thoracic and abdominal cavities. It has four qualities like glossopharyngeal nerve (IX):

- somatomotor control: parts of the pharynx and larynx muscles (speaking).
  Nucleus: Nucleus ambiguus.
- somatosensory control: larynx, parts of the external ear.
  Nucleus: spinal nucleus of the trigeminal nerve.
- visceromotor control: parasympathetic innervation of bronchi, bronchioli, atria of the heart, liver, gallbladder, pancreas, oesophagus, stomach, gut (as far as colon flexure) and kidneys.
  Nucleus: dorsal nucleus of the vagus nerve.
- viscerosensory: digestive tract, lungs, heart and aortic arch.
  Nucleus: solitary nucleus.

Skull: jugular foramen.

Vallecula cerebelli

Meninges & Cisterns   1

TA Latin: *Vallecula cerebelli*
TA English: *Vallecula of cerebellum*

The valecula cerebelli is the name given to the valley on the ventral side of the cerebellum, arising between the tonsils of cerebellum and having its depth curtailed by the uvula vermis.

Varolius bridge

TA Latin: *Pons (Varolius)*
TA English: *Pons of Varolius*
→ Pons (Varolius)

Vein of pontomedullary sulcus

Vessels   1

Inconstantly manifest vein.

Vein of pontomesencephalic sulcus

Vessels

TA Latin: *V. pontomesencephalica*
TA English: *Pontomesencephalic vein*
→ Anterior pontomesencephalic vein

Vein of the medulla oblongata (anterolateral)

Vessels   2

Inconstant veins coursing along the myelencephalon and draining it.

Vein of the olfactory gyrus

Vessels   1

TA Latin: *V. gyri olfactorii*
TA English: *Vein of olfactory gyrus*
Superficial vein draining the olfactory gyrus.

Veins of encephalic trunk

Vessels   1

TA Latin: *Vv. trunci encephali*
TA English: *Veins of brainstem*
Veins of the brainstem.

Veins of the caudate nucleus

Vessels   2

TA Latin: *Vv. nuclei caudati*
TA English: *Veins of caudate nucleus*
Several veins drain venous blood from the caudate nucleus into the internal cerebral vein,

from which it passes via the great cerebral vein and the straight sinus to the confluence of the sinuses.

Venous plexus of oval foramen

Vessels   2
TA Latin: *Plexus venosus foraminis ovalis*
TA English: *Venous plexus of foramen ovale*
Venous plexus in and on the oval foramen, which is connected with the cavernous sinus. From here, it goes to the sigmoid sinus via the superior petrosal sinus.

Ventral amygdalofugal fibers

Pathways   1
The ventral tract coming from the amygdaloid body contains fibers of the basolateral group going in the direction of the hypothalamus, medial thalamic nuclei, septal verum and cerebral cortex.

Ventral anterior thalamic nucleus

Diencephalon   3
TA Latin: *Nucl. vent. ant. thalami*
TA English: *Ventral anterior nucleus of thalamus*
The ventral anterior (VA) nucleus and ventral lateral (VL) nucleus are involved in somatomotor integration. Their afferents come from the cerebellum and basal ganglia. Their efferents go to the premotor cortex as well as to the precentral gyrus (area 4). The convergence of cerebellar and basaloganglionic signals attests to the important role played in processing and initiation of voluntary motor activities.

Ventral ascending serotoninergic tract

Pathways   1
This very large bundle comes primarily from the dorsal raphe nucleus and from the superior central nucleus.
The tract dispatches fibers to virtually all regions of the midbrain and diencephalon, as well as reaching cortical centers.

Ventral cochlear nucleus

Myelencephalon   3
TA Latin: *Nucl. cochlearis ant.*
TA English: *Anterior cochlear nucleus*

→ Cochlear nuclei

Ventral lateral thalamic nucleus

Diencephalon   3
TA Latin: *Nucl. vent. lat. thalami*
TA English: *Ventral lateral nucleus of thalamus*
→ Ventral anterior thalamic nucleus

Ventral posterior thalamic nucleus

Diencephalon   2
TA Latin: *Nucl. ventrobasales thalami*
TA English: *Ventrobasal complex of thalamus*
The ventral posterior (VP) nucleus is involved in somatosensory integration. Its medial segments (VPM) receive the somatotopically arranged trigeminothalamic projections, whereas the lateral segment (VPL) receives the spinothalamic projections. Epicritic and protopathic signals of the contralateral body half are processed. Projections go to the postcentral gyrus and surrounding areas.

Ventral posterolateral thalamic nucleus

Diencephalon   2
TA Latin: *Nucl. vent. posterolat. thalami*
TA English: *Ventral posterolateral nucleus of thalamus*
This lateral part of the ventral posterior thalamic nucleus (VPM) is the final station of the trigeminothalamic projections, conducting epicritic, protopathic and gustatory information primarily from the contralateral but also the ipsilateral body halves. Afferents and nucleus are somatotopically organized. Projections go to the postcentral gyrus and insula.

Ventral posteromedial nucleus, parvocellular part

Diencephalon   2
TA Latin: *Nucl. vent. Posteromed., pars parvocellularis*
TA English: *Ventral posteromedial nucleus, parvocellular part*
The medial, parvocellular portion of the ventral posteromedial thalamic nucleus contains afferents from the ipsilateral trigeminothalamic tract, amygdaloid body, tegmentum area, gustatory neocortex and medial parabrachial nucleus and conveys gustatory information to the insula.

Thus this region plays an important role in the integrative processing of gustatory information.

Ventral posteromedial thalamic nucleus
Diencephalon   2
TA Latin: *Nucl. vent. posteromed. thalami*
TA English: *Ventral posteromedial nucleus of thalamus*
This medial part of the ventral posterior thalamic nucleus is also abbreviated VPM and is the final station of the trigeminothalamic projections, conducting epicritic, protopathic and gustatory information primarily from the contralateral but also the ipsilateral body halves. Afferents and nucleus are somatotopically organized. Projections go to the postcentral gyrus and insula.

Ventral premammillary nucleus
Diencephalon   1
TA Latin: *Nucl. premammillaris vent.*
TA English: *Ventral premammillary nucleus*
→ Dorsal premammillary nucleus

Ventral root of the spinal nerve
Medulla spinalis   3
TA Latin: *N. spinalis, radix ant.*
TA English: *Anterior root of the spinal nerve*
Motor fibers emerge from the spinal cord to the periphery in the ventral root of a spinal nerve.

Ventral striatum
Telencephalon   1
TA Latin: *Striatum ventrale*
TA English: *Ventral striatum*
Ventral part of the corpus striatum.

Ventral tegmental area (Tsai)
Mesencephalon   1
TA Latin: *Nucl. subbrachialis (Tsai)*
TA English: *Subbrachial nucleus (Tsai)*
Apart from the cells of the substantia nigra, compact part, the neurons of this area also produce dopamine. The area is situated in the tegmentum of mesencephalon, directly beside the likewise dopaminergic pigmental parabrachial nucleus. Both nuclei can be poorly differentiated from the substantia nigra. The ventral tegmental area belongs to the limbic region

of the midbrain and is involved in affective motor processes and perception of thirst.

Ventral tegmental decussation (mesencephalon) (Forel)
Mesencephalon   1
TA Latin: *Decussatio tegmentalis ant. (Forel)*
TA English: *Anterior tegmental decussation (Forel)*
In the ventral tegmental decussation the fibers of the red nucleus decussate to the contralateral side and pass as the rubrospinal tract into the cervical cord.

Ventral tegmental fasciculus
Mesencephalon   2
These fibers are efferent fibers of the principal nucleus of the trigeminal nerve. After exiting from the nucleus, these decussate to the contralateral side (ventral tegmental fasciculus), bundle there to form the trigeminal lemniscus and then ascend to the ventral posteromedial thalamic nucleus. They conduct somatosensory information from the entire facial surface.

Ventral tegmental nucleus
TA Latin: *Nucl. tegmentalis ant.*
TA English: *Anterior tegmental nucleus*
→ Ventral tegmental area (Tsai)

Ventral thalamic nuclei
Diencephalon   3
TA Latin: *Nuclei ventrales thalami*
TA English: *Ventral nuclei of thalamus*
The ventral nuclear group of the thalamus is composed of four nuclei. Whereas the ventral anterior nucleus and ventral lateral nucleus are integrated in the somatomotor control system, the ventral posterior nucleus is the main thoroughfare for all somatosensory information on its way to the cerebral cortex. The ventral posterolateral thalamic nucleus conveys protopathic and epicritic information from the trigeminal complex, the ventral posteromedial nucleus from the spinal cord.

Ventricular system
Meninges & Cisterns   3

TA Latin: *Ventrikelsystem*
TA English: *Ventricular system*
Intraneural liquor space. Connected with the extraneural liquor space, the cisterns, via aperturae (mediana + lat.) ventriculi quarti.

Ventrolateral superficial reticular area

Pons   2

Poorly delineated region of the lateral reticular formation which corresponds rostrally with the lateral gigantocellular reticular nucleus and caudally with the retroambiguous nucleus. Functionally it is, however, one entity which plays a role in cardiovascular and respiratory regulation as well as in pain suppression. It has reciprocal connections with the spinal cord and hypothalamus.

Ventromedial hypothalamic nucleus

Diencephalon   2

TA Latin: *Nucl. ventromed. hypothalami*
TA English: *Ventromedial nucleus of hypothalamus*
Hypothalamic nucleus that plays a decisive role in combat and satiation behavior (lesions mediate extreme anger outbursts, hyperphagia and obesity). Direct and indirect afferents from the subiculum and amygdaloid body, peripeduncular nucleus, lateral parabrachial nucleus as well as many other centers. Efferents go to the periaqueductal gray and to the reticular formation of mesencephalon.

Venula

Vessels   1

TA Latin: *Venula*
TA English: *Venule*
Small venula that drains the surrounding tissue.

Vermis cerebelli

Cerebellum   3

TA Latin: *Vermis cerebelli (I-X)*
TA English: *Vermis of cerebellum (I-X)*
Vermis cerebelli is the name given to the entire middle region between the two cerebellar hemispheres. It receives its afferents from the spinocerebellar tracts and is thus also called the spinocerebellum. Some of its efferents (from zone A) course via the fastigial nucleus to the thalamus and to the vestibular nucleus. Another

part (from zone B) passes without synapsing to the lateral vestibular nuclei.

Vermis cerebelli, A zone

Cerebellum   1

A zone denotes the part of the vermis cerebelli projecting to the fastigial nucleus and from here to the vestibular nuclei and thalamus. This zone accounts for the larger portion of the vermis. The B zone conversely dispatches its efferents directly without synapsing to the lateral vestibular nucleus.

Vermis cerebelli, B zone

Cerebellum   1

→ Vermis cerebelli, A zone

Vertebral artery

Vessels   3

TA Latin: *A. vertebralis*
TA English: *Vertebral artery*
The vertebral artery arises from the subclavian artery (prevertebral part). At the level of C6, it enters the transverse process foramen and passes to C2 (transverse part). Here it forms a large arch before entering the transverse process foramen of the atlas. In this manner, it can easily track head movements (vertebral artery, atlantal part). Via the foramen magnum it passes into the cranial cavity, where the right and left arteries unite at the lower margin of the pons and form the basilar artery (intracranial part).

Vertebral artery, atlantal part

Vessels   2

TA Latin: *A. vertebralis, pars atlantica*
TA English: *Vertebral artery, atlantic part*
Before entering the transverse process foramen of the atlas, the vertebral artery describes an outwards-curved arch (siphon) which permits it to follow head movements without extension or compression. After penetrating the atlas, the artery courses to the foramen magnum where it enters the cranial cavity.

Vertebral artery, intracranial part

Vessels   2

TA Latin: *A. vertebralis, pars intracranialis*
TA English: *Vertebral artery, intracranial part*

Once the vertebral artery has entered the cranial cavity through the foramen magnum, it is refered to as the intracranial part. Directly behind the foramen, the meningeal branches are given off to supply the calvaria and cranial meninges. On its further course, branches are given off: posterior spinal artery, olivary arteries, posterior inferior cerebellar artery and anterior spinal artery.

The intracranial part passes as far as the transition between bulb and pons, where the paired vessels unite to form the basilar artery.

Vertebral artery, intracranial part, posterior meningeal branch

Vessels   1
TA Latin: *A. vertebralis, pars intracranialis, R. meningeus post.*
TA English: *Vertebral artery, intracranial part, posterior meningeal branch*
Immediately after entering the cranial cavity through the foramen magnum, the anterior meningeal branch and the posterior meningeal branch are given off, supplying bones and cranial meninges.

Vertebral artery, spinal branch

Vessels   1
TA Latin: *A. vertebralis, R. spinalis*
TA English: *Vertebral artery, spinal branch*
Lateral branches of the vertebral artery, transverse part. The branches pass to the spinal nerves and supply spinal ganglia, spinal cord and spinal meninges.

Vertebral artery, transverse part

Vessels   1
TA Latin: *A. vertebralis, pars transversaria*
TA English: *Vertebral artery, cervical part*
Transverse part is the term used to designate the portion of the vertebral artery which passes through the transverse process foramina of vertebrae C6 to C2 in the direction of the skull. The vertebral artery, atlantal part begins at C1.

Vertebral vein

Vessels   3
TA Latin: *V. vertebralis*
TA English: *Vertebral vein*

The vertebral vein is an artery that runs through the transverse foramina of the 1st to 6th (7) cervical vertebrae. It flows into the ipsilateral brachiocephalic vein.

Vestibular area

Pons   1
TA Latin: *Area vestibularis*
TA English: *Vestibular area*
Area of the fourth ventricle lying above the vestibular nuclei.

Vestibular ganglion

Ear   3
TA Latin: *Ganglion vestibulare*
TA English: *Vestibular ganglion*
In the vestibular ganglion are situated the somas of the bipolar ganglion cells which receive the signals from the sensory cells and conduct them in the direction of the brainstem. In doing so, they project to the vestibular nuclei (medial, superior, inferior) and to the cerebellum.

Vestibular nerve

Nerves   3
TA Latin: *N. vestibularis*
TA English: *Vestibular nerve*
Part of cranial nerve VIII, the vestibulocochlear nerve.
Predominantly sensory nerve whose fibers convey information from the vestibular parts of the inner ear (sacculi, utricles and semicircular canals) to the vestibular nuclei (medial, superior and inferior) and the cerebellum. The somas of the bipolar ganglion cells are located in the vestibular ganglion.

Vestibular nerve, afferent fibers

Nerves   2
The vestibular nerve contains mostly afferent fibers to the vestibular nuclei (medial, superior and inferior), but also efferent fibers which pass on to the sensory cells and exert an inhibitory effect on these.

Vestibular nerve, descending branch

Nerves   1
After entry into the brainstem, the vestibular nerve divides into one ascending branch and one descending branch.

Vestibular nerve, efferent fibers

Nerves   2
→ Vestibular nerve, afferent fibers

Vestibular nuclei

Pons   3
TA Latin: *Nuclei vestibulares*
TA English: *Vestibular nuclei*
The vestibular nuclei are the equilibrium component (vestibular part) of vestibulocochlear nerve (VIII). The following are differentiated:
– superior vestibular nucleus,
– lateral vestibular nucleus (Deiters),
– medial vestibular nucleus,
– inferior vestibular nucleus.
They have intensive connections with the labyrinth, cerebellum, spinal cord, motor cranial nuclei as well as the cerebral cortex and are actively involved in manifold balance and orientation tasks.

Vestibular nuclei (medial, superior , inferior)

Pons   3
TA Latin: *Nuclei vestibulares (med., sup., inf.)*
TA English: *Vestibular nuclei (medial, superior, inferior)*
The three vestibular nuclei are the actual projection area of the vestibular ganglion, whereas the lateral vestibular nucleus (Deiters) can be viewed almost as an outpost of the cerebellum.
Afferents: ampulla → superior + medial,
sacculi and utricles →medial + caudal.
Cerebellar efferents come from the flocculus, nodulus and fastigial nuclei.
Efferents to the cerebellum, spinal cord, cerebral cortex and motor cranial nerves.

Vestibular system

General CNS   3
→ Vestibular nuclei

Vestibulo-ocular reflex path

Pathways   1
Reflex coupling of eye control for body and head movements.

Vestibulocerebellar fibers, primary

Pathways   1
Fibers projecting from the vestibular nuclei to the cerebellum, without synapsing.

Vestibulocerebellar fibers, secondary

Pathways   1
Fibers projecting from the vestibular nuclei to the cerebellum, with synapsing.

Vestibulocerebellar tract

Cerebellum   2
The vestibulocerebellar tract is the global term for all afferents coming directly from the vestibular organ or the vestibular nuclei and going to the cerebellum.
As mossy fibers, they pass here into the flocculonodular lobe, which for this reason is also called the vestibulocerebellum. Collaterals course to the fastigial nucleus whose afferents are returned to the vestibular nucleus, thus giving rise to a feedback loop.

Vestibulocochlear nerve (VIII)

Nerves   3
TA Latin: *N. vestibulocochlearis (N.VIII)*
TA English: *Vestibulocochlear nerve (VIII)*
Vestibulocochlear nerve (VIII) is composed of two parts:
– vestibular nerve: it is responsible for innervation of the vestibular structures of the inner ear (sacculi, utricles and semicircular canals).
  Nucleus: vestibular nuclei.
– cochlear nerve: it innervates the cochlea and is the first element of the auditory tract.
  Nucleus: cochlear nuclei.
Skull: internal acoustic meatus.

Vestibulomesencephalic tract

Pons   1
Fibers going from the vestibular nuclei in the direction of the mesencephalon. They originate in the medial vestibular nucleus and superior vestibular nucleus, projecting to nuclei of the extraocular muscles, i.e. the abducens nucleus and nucleus of the trochlear nerve the oculomotor nucleus, interstitial nucleus of the stria terminalis (Cajal) and rostral interstitial nucleus of the medial longitudinal fasciculus.

Vestibulothalamic fibers

Pathways
→ Vestibulothalamic tract

Vestibulothalamic tract
Diencephalon 2
Fibers from the vestibular complex run in the vestibulothalamic tract to the ventral nuclear group of the contralateral thalamus.

Vicq d'Azyr bundle
Pathways 1
→ Mammillothalamic fasciculus, Vicq d'Azyr bundle

"Visceral cortex"
Telencephalon
TA Latin: *Operculum*
TA English: *Operculum*
→ Frontoparietal operculum

Viscero-afferent fibers
Medulla spinalis 1
Information from the autonomic nervous system (glands, smooth musculature etc.)

Viscero-efferent fibers
Medulla spinalis 1
Effector information in the direction of the autonomic nervous system (glands, smooth musculature etc.)

Visual cortex, primary
Telencephalon
→ Area 17 (striate cortex)

Visual cortex, secondary
Telencephalon 3
Corresponds to area 18 and area 19. The visual signals that had been the target of preliminary processing in area 17 are subjected to further associative processing.

Visual field
→ Field of vision, binocular part

WXYZ

Wallenberg

Myelencephalon

TA Latin: *Tractus trigeminothalamicus post. (Wallenberg)*

TA English: *Posterior trigeminothalamic tract (Wallenberg)*

→ Dorsal trigeminothalamic tract (Wallenberg)

Wernicke speech center

Telencephalon   3

The superior temporal gyrus lies at the upper margin of the temporal lobe.

It comprises Broadmann areas 42 and 22, which together form the secondary auditory cortex, i.e. Wernicke's area. Here tones are processed associatively, hence interpreted, relativized and combined with memory contents.

White commissure (of spinal column)

Medulla spinalis   2

TA Latin: *Commissura alba (medullae spinalis)*

TA English: *White commissure (of central cord)*

Bridge of white matter at the end of the anterior fissure.

Here run commissural fibers connecting the anterior funiculi (anterior columns) of both sides.

White laminae of the cerebellum

Cerebellum   1

The white lamina of the cerebellum is the term used to designate the white, lamella-like extensions from the medullary body of cerebellum. These consist of fibers ascending to and descending from the cerebellar cortex, and forming the arbor vitae that can be recognized in the median section.

Wing of central lobule

Cerebellum   2

TA Latin: *Ala lobuli centralis*

TA English: *Wing of central lobule*

Wing-shaped extension from the central lobule.

Wing of cisterna ambiens

Meninges & Cisterns   1

Lateral extension from the cisterna ambiens.

Zona incerta

Diencephalon   2

TA Latin: *Zona incerta*

TA English: *Zona incerta*

The zona incerta belongs to the subthalamus and is composed of dispersed small cells, so that it is viewed as being a continuation of the reticular formation. It has no cortical efferents but rather projects to the tectum, tegmentum of mesencephalon, raphe nuclei and motor cranial nerve nuclei.

Afferents have their origin in the pyramidal tract, cerebellum and globus pallidus.

Zonal layer of the superior colliculus

Mesencephalon

TA Latin: *Stratum zonale colliculi sup.*

TA English: *Zonal layer of superior colliculus*

→ Superior colliculus, zonal layer